AF781989

new technologies
in global societies

new technologies in global societies

editors

pui-lam law

The Hong Kong Polytechnic University, China

leopoldina fortunati

University of Udine, Italy

shanhua yang

Peking University, China

World Scientific

NEW JERSEY • LONDON • SINGAPORE • BEIJING • SHANGHAI • HONG KONG • TAIPEI • CHENNAI

Published by

World Scientific Publishing Co. Pte. Ltd.

5 Toh Tuck Link, Singapore 596224

USA office: 27 Warren Street, Suite 401-402, Hackensack, NJ 07601

UK office: 57 Shelton Street, Covent Garden, London WC2H 9HE

Library of Congress Cataloging-in-Publication Data

New technologies in global societies / edited by Pui-Lam Law, Leopoldina Fortunati & Shanhua Yang.

p. cm.

Includes bibliographical references and index.

ISBN 981-256-812-3

1. Technological innovations. 2. Technology. 3. Globalization. I. Luo, Peilin. II. Fortunati, Leopoldina. III. Yang, Shanhua.

T173.8.N477 2006

303.48'3--dc22

2006044632

British Library Cataloguing-in-Publication Data

A catalogue record for this book is available from the British Library.

Typeset by Stallion Press

Email: enquiries@stallionpress.com

Printed in Singapore by World Scientific Printers (S) Pte Ltd

Acknowledgments

Most of the papers in this collection emerged from the session on "The co-construction of society and technology in a global world," at the 36th World Congress of the International Institute of Sociology held in Beijing on July 7–12, 2004. We would like to thank the authors for their relentless commitment and their contributions to this collection. We would also like to express our gratitude to the reviewers for their thoughtful and meticulous comments. Finally, our thanks go to the Department of Applied Social Sciences of The Hong Kong Polytechnic University for their support of the publication of this book.

Introduction

Pui-lam Law, Leopoldina Fortunati and Shanhua Yang

Technological advancements since the last millennium in the West have contributed to global modernity. They have led to the development of global connectivity (Wellman *et al.* 2005), mobility, and communication, and in turn, have occasioned globalization (Tomlinson 1999). Aviation technology has shrunk global spatial distance with the result that international mobility is more convenient and countries that are thousands of miles apart can now be reached in just a number of hours. Thus, air travel has been conducive to global closeness. The development of print, radio, television, the telephone, and the Internet have made global communication possible and people now have more access to the world (Barnett *et al.* 2003). Through these communication technologies people are able to instantly understand what is happening in other countries. For example the news of the July 7th, 2005 bombings in London was immediately transmitted all over the world and people had *real time* knowledge of what had happened. In addition, people are no longer restricted to body-to-body interactions within their own localities; they are able to connect with people from different localities throughout the world by fixed line or mobile phones or the Internet (Nyiri 2005).

Through the process of globalization the local is inevitably connected with the global, and the mix of global and local has restructured the social lives of the locals (Strassoldo 1992). For instance, an increase of just 2% on the average exchange rate between the Chinese yuan and the U.S. dollars has led to the closing of several small factories in a township in the southern part of China and in turn influenced the lives of migrant workers there, with family members in other parts of China being affected accordingly. People have started to be aware of the fact that when they make plans for their future knowledge of the local is insufficient without information of the global. Social relations have become more global as they have been influenced by processes or events that have taken place at a distance. Thus, some have argued that the local will in time be displaced by the global and consequently a global culture will be generated (Appadurai 1997).

Despite the fact that nation-states are globally connected, the emergence of a global culture is highly unlikely. It is widely believed that global modernity is indeed Western modernity, and that the powerful force of Western modernity is the sweeping expansion of its market economy (Sen 1999; Appadurai 1990). With the advancement of technology which makes connectivity and mobility among people of different nation-states more efficient and effective, the market economy has expanded its influence all over the world (Braman 1990). Yet Latouche (1996) clearly points out that the expansion of the Western style market economy, powerful though its impact is, is not able to uproot the different cultures of the so-called third world. Thus, it is an untenable belief that the adoption of a market economy will bring about the changes at social and cultural levels in the third world or non-Western societies. The development of the market economy in the West indeed has its own historical, social and cultural contexts which are different from non-Western societies (Gray 1997; Taylor 1999; Wallestein 1974). On the contrary, without the respective cultural elements of Western societies, a market economy can barely root itself in a proper fashion in non-Western soil. Indeed, economics in the East are specifically Asian in style.

The cultures of non-Western countries in fact have their own countervailing forces that counteract the power of the global economy. Thus this suggests that the powerful disembedding force of the global market economy can hardly generate a global culture, let alone a global society. It is widely agreed that the nation-states are well connected and consequently globalized, but the dialectical process of the global and local generates global *societies*.

The inherent connectedness of the social, cultural, political, and economic systems of society which give rise to the dialectical process of the global and local also illuminates an understanding of science and technology and their relationship with global societies. Science and technology have rapidly developed over the past three centuries in Western countries. Science has also presented itself in the form of universal knowledge. Ostensibly, science is in itself an autonomous intellectual enterprise aimed at understanding nature. Yet the flourishing of science in the West is intrinsically related to the socio-cultural changes that took place during the age of Enlightenment; and technology, the instrumental application of science, is without doubt inextricably linked to social needs (Maldonado 2005; Longo 2002). The shift in the technological paradigm is therefore, imputed more to changes in social needs and less to the result of the rational discussion of the autonomous scientific community.

Taking the social nature of science and technology into account in order to understand their effects on global societies is essential. It is evident that science and technology are globally oriented and that their rapid development have resulted in global modernity. Technologies set conditions for the closeness of the nation-states and for the affinity of the global and the local. They are also penetrating everyday life, and even sometimes the body, producing radical social changes. For instance, Chen in this volume (Chapter 9) demonstrates that Internet use, which enables Chinese immigrants to Canada to engage in transnational entrepreneurship, shows the positive function of new technology. Yet, arguing that new technologies bring new life and a promising future to global societies remains a questionable thesis. For example, Pertierra (Chapter 2) succinctly points out that various aspects of everyday culture in Philippine society inhibit an orientation to science and that their undeveloped economy is unable to sustain high-level skills in technology. All these would prevent a scientific ethos from establishing local roots in the Philippines. In addition, Law and Peng (Chapter 11) maintain that, although the mobile phone is a new personalized communications device, its use among migrant workers in southern China, is one of the factors which revitalizes traditional kinship networks instead of bringing about a more individual lifestyle. More negatively, Liu and Lau (Chapter 12) have demonstrated that the Internet, which serves as an advertisement tool for the sex trade, has occasioned a new form of domination of sex workers through cyberspace.

This collection attempts to explore the relationship between new technologies and global societies, to gain an understanding of how the positive as well as negative influences of technologies bear on global societies, how their practices of use are resisted or re-interpreted by them, and how their social meaning is constituted through the process of negotiation with these societies. Part 1 is on science, technology, culture, and the body; Part 2 on new media and generations; and Part 3 on information and communication technologies (ICTs) and work.

Part 1 Technology, Science, Culture, and the Body

In Chapter 1, Pertierra discusses the cultural factors of everyday life that affect the development, practice and valuation of science in Philippine society. Despite various aspects that should encourage the development of science, such as widespread educational opportunities, an orientation towards Western modernity and the establishment of scientific institutions early in the American period, science in the Philippines has not generally prospered. Pertierra maintains that there are complex reasons for this relative lack of scientific development among which are a weak economy and a political system that neglects investments in long-term projects such as science education. There are also cultural elements of everyday life that prevent the development of a scientific ethos, such as the weak development of the public sphere, an undervaluing of specialized skills and an animated view of nature.

Prior work on university-industry relations has mostly focused on developed countries such as the United States, Japan and the countries of Western Europe, examining the shift in science policy from curiosity-driven research (Mode 1) to user-driven research (Mode 2). Little research has been done in China, where a Mode 2-oriented policy has been persistently advocated since the 1950s. In the second chapter, Hong describes various forms of technology transfer from Chinese universities to industry that she identified through 40 in-depth interviews with university scientists and technology transfer officers in China. While a negative impact of commercial behavior on scientific research was found in low-level university-industry collaborations, scientists involved in higher-level industry-related research benefited from enhanced communication with colleagues. In contrast with the extensive interaction between applied research and development, the connection between basic and applied research is quite weak in China. Hong warns that, although basic research is not confined by a country's borders, knowledge transfer is. Moreover, low absorptive capacity due to underinvestment in basic science will preclude China from sharing the most advanced scientific knowledge. A Mode 2-oriented science policy, though seemingly practical in the short run, will impair China's long-term international competitiveness.

There has long been discussion on the artificialization of the human body (Katz 2002; Fortunati 2002). People are increasingly confronted with new technologies getting closer to the human body, and the penetration of the latter by the former. The aim of this research was to monitor how the human body and new technologies, mainly the mobile and the Internet are perceived in this critical phase termed "mass prosthetization." By utilizing a social representations approach (Moscovici 1961/76; Jodelet 1989) as their framework, Contarello and Fortunati, in Chapter 3, explore the forms of social co-construction of the "technological" body being worked out in daily life. Their chapter is based on a study which explored symbolic and emotional levels relating to the human body, mobiles and the Internet, and analyzed bodily experience and attitudes towards mobiles and the Internet, gathered information on practices relative to these technologies, and finally examined opinions on the social impact of these on the human body. More than 200 Italian university students were interviewed and data was analyzed both through qualitative and quantitative procedures. In their chapter Contarello and Fortunati indicate that there were no straightforward rejections, nor great enthusiasm concerning the invasion of the body by machines, though a clear opposition between the human body and technology is emerging. In addition, a positive attitude towards technologies is confirmed, although with some concern. The Internet is perceived as having a twofold nature: work and play, hot and cold, science and chaos. Mobiles are associated with a rising "price-to-be-paid."

Technology can only develop within a social system. If one imagines a global world, technological development would be quite similar in all countries. The example of Preimplantation Genetic Diagnostics (PGD) shows that the German government chooses not to allow this technology, while it is legal in most other European countries. Wüstner in Chapter 4 analyzes why this is the case, and shows how the discourse about PGD in Germany can be described. In order to investigate it, Wüstner carried out a content analysis of several German newspapers and magazines and elaborated typical lines of argument. The content analysis focused on how processes of social construction can be reconstructed and how communication and legitimation processes of the structure and pattern of sense can be explained. The analysis presented in Chapter 4 results in the supposition that it is not *the* society that decides whether PGD is allowed or forbidden but that there are relevant social subsystems that are of importance in this respect. She further concludes that, although the relationship between technology and society is co-constructed, understanding how technology is socially represented within different social systems is also important as this is a crucial element of social thinking.

Part 2 New Media and Generations

With the increasing popularity of the Internet around the globe, newsgroup communication has acquired great significance as a source of documentation in social research. Yet the scanning of the contents of messages is difficult to perform within traditional reading and classification techniques, because of the huge amount of information and "background noise" (irrelevant elements) that interfere with the selection of significant "markers." Chapter 5 is a research report which aims to probe the potentiality of the technique called "the automatic analysis of textual data" (textual mining), in order to scan the content of newsgroups' messages. The objective is to minimize the loss of information on one hand, and to increase the meaningfulness of thematic inspection on the other, to serve as the possible starting point for a qualitative analysis. Research was carried out by Giuliano on eight Usenet Italian newsgroups from March 18 to April 18, 2005, during the Iraq war. The corpus of the research was made up of 35,492 messages, out of a total of 5,220,932 occurrences. Two text-*mining* software programs were used: TALTAC, for the normalization, grammar tagging, and stemming of the Italian language; and SPAD-T 1.5/SPAD 5.0 for the multivariate analysis of textual data. The results confirm the overall performance potentiality of the strategy used in the selection of key words that properly describe the contents of messages. The results of the research show that in the newsgroups that were examined, there was general opposition to the war; however, different reasons were given and different viewpoints expressed. As a caveat, we must consider that it is not always possible to identify the ideological orientation of newsgroups because users seek to engage in debate and polemical confrontations, and may also express thoughts that go beyond their own political beliefs.

With the tremendous information flow through new communication technologies which is supposed to influence people's ideological orientation, such as attitudes towards the Iraq war, Fortunati and Strassoldo in Chapter 6 investigate how the process of negotiation between societies and information technologies has been evolving in the global world, producing important changes in the organization of everyday life as well as in the widespread production of attitudes and opinions towards relevant social themes. They focus in particular on exploring relations between ICTs and youth in the age of globalization in Italy. They explore on one hand the use and quantity of use of ICTs, like TV, computers, and the use of more classical cultural instruments, like books and magazines, among youths of both genders and on the other hand, the diffusion and degree of diffusion of modalities of communicative sociality, with the aim of verifying if and on what terms ICT use correlates with practices of communicative sociality. Secondly they investigate how these youths see socio-political themes like one's own country, peace, war, terrorism, military service, and how they judged the events of Genoa where anti-globalization protesters clashed with police during a G8 meeting. In this chapter they also try to verify if and how the attitudes of youths towards all these themes correlate with their use of ICTs and their modes of communicative sociality. This analysis is particularly important for verification of the hypothesis that the existence of a media regime in Italy, undeniably controlled by the premier Berlusconi, has a strong conditioning effect on public opinion, and particularly youth opinion. Their analysis is based on a representative sample of Italian youth, interviewed by means of a questionnaire conducted face to face in June 2002. The data was analyzed by means of multivariate analysis methods.

Research into teenagers and mobile phone use reveals that parents purchase their teenage children mobile phones to keep tabs on them and monitor their movements; another form of what Rakow and Navarro (1993) have dubbed "remote mothering." Sciriha in Chapter 7 discusses the results of a large-scale survey on mobile telephony, conducted in November 2003 among a scientifically representative sample of 500 respondents in Malta, a small island in the center of the Mediterranean. The survey sought to answer a number of questions including; when compared to Maltese in the older age groups, what role do teenagers play in the high penetration levels of mobile phones? What reasons do they give for phoning on mobiles rather than on fixed lines, even though mobile phone rates are very expensive compared with fixed lines? How long are teenagers willing to wait for answers to their urgent and/or not so urgent SMS messages? Do they lie when they are asked "Where are you?" From the results of the survey, Sciriha found that teenagers are the most enthusiastic users of mobile phones, and that mobile phones provide them with an earlier-than-anticipated independence from their parents. Although mobiles have provided parents with a new means of caring for their young, the sense of being able to be in immediate contact can be real and unreal. According to the data teenagers lie to their parents when contacted on their mobiles. She further maintains from her findings that telephones, especially mobile phones afford privacy to their users. They have also contributed to giving people a sense of freedom when in reality they have managed to enslave them. Once introduced, they cannot be ignored and the demand for them is still on the increase.

Mobile phones have also provided a new means for caring for the aged. Wong in Chapter 8 explores the use of mobile phones among residents of care homes for the elderly. This chapter examines the family relationships and communication patterns between residents of care homes and their family members with regard to the use of mobile phones. Wong introduces his research by observing that the changing structure of Hong Kong society has had an impact on the structure of the family. Families are shrinking in size, and if support networks for the elderly are available, they tend to be limited, which explains the development of care homes for the elderly. Data for the research was collected from a survey based on two sets of interview schedules. Mobile phone users who were resident in care homes were interviewed, and information on the home's regulations on the use of mobile phones was also collected from the home management. According to the findings, there are more male mobile phone users than female users, and women are slightly more likely than men to be daily users. The use of mobile phones is more about safety and security with some of the respondents stating that they feel better when they use their mobile phones. Wong concludes that the mobile phone is able to create a "cyber space" between the generations and that it promotes family resilience since spoken words of care count.

Part 3 ICTs and Work

At the intersection of the two literatures — immigration transnationalism and the interaction of technology and society — the role of Internet use with regard to transnational entrepreneurship provides a unique vantage point from which to observe the ways in which social, cultural, and technological forces affect economic actions. The impact of the Internet on transnationalism is frequently mentioned yet rarely researched. In Chapter 9 Chen attempts to explore the socio-technological dimension of transnational entrepreneurship by examining how Chinese immigrants to Canada rely on communication and social networks to engage in transnational entrepreneurship. Participant observations and in-depth interviews conducted in Toronto and Beijing indicate that Internet use has positive effects on transnational entrepreneurship in terms of searching for information, building and cultivating border-crossing social networks, and mobilizing important resources. However, the Internet is not everything, as transnational entrepreneurs selectively use multiple forms of communication media, depending on which is most convenient and socially appropriate.

In Chapter 10 Chu and Yang try to depict the push-pull factor between ICTs as signifiers of the age of globalization on the one hand, and the reinforcement and/or distortion of cultural factors for contemporary Chinese on the other. By means of intensive case studies on young peasant migrant workers working in townships in a South China village, the two researchers were able to examine the impact of social changes on the normative basis of this group of people. Chu and Yang tried to comprehend how their respondents perceived interpersonal relationships, re-interpreted kinship ties when distance was an issue, and how they interpreted the relationship between the individual and society with reference to their experience of living in a global era. By studying their patterns of mobile phone use, including text messaging, the duration of conversations, the selection of user packages, and the criteria for choosing phone models, the researchers were able to articulate what had and had not changed about peasant identity vis-à-vis Chinese cultural expectations, and the possible societal transformation within rural China.

Law and Peng in Chapter 11 echo the observations made by Chu and Yang by describing mobile phone consumption patterns of migrant workers and how mobile phones affect their social relationships. Law and Peng maintain that mobiles provide conditions for the displacement of traditions. Their interviews have shown that migrant workers are able to develop virtual connections with strangers and are able to join virtual networks through their mobile phones since these kinds of cyber relationships have the potential to provide a release from the problems of the real world. Yet unfavorable social conditions of the host society have also re-anchored them to networks within their places of origin. The mobile phone has made it possible for workers to reconnect either in person or through cyberspace, with their kinsmen or fellow villagers who are scattered far and wide throughout Guangdong, or even in other provinces. Wherever these kinsmen and villagers are, traditional village norms exist in the form of a floating cyber network (Gergen 2002, 2003) that exercises constraints on them when they are connected by mobile phone, either by SMS or by phone calls. Interestingly enough, the mobile, as a new kind of communication technology, has re-embedded migrant workers within the bosom of tradition.

Sex work is one of the oldest forms of work in human societies and one of the most crucial problems at both the social and political level (Fortunati 1981, Amer. Trans. 1995). The phenomenal development of information technology over the past decade, however, has affected this work in a new way, altering the conventional mode of disseminating sex information. In Chapter 12 Liu and Lau set out to investigate how the proliferation of sex service information in Hong Kong, particularly concerning "one-woman brothels," is affected by the Internet. They do so by examining the contents of relevant websites identified via the search engines Google and Yahoo, as well as those under other relevant local web directories. The focus of the chapter is on the kinds of information available on sex services, on how this information is presented, and on how far this mode of information differs from that of conventional media like advertisements in newspapers and magazines. By examining these issues, Liu and Lau try to clearly depict the mode of presentation of sex service information on the web, showing that the power of sex workers is diminishing because of the further level of control and conditioning represented by the technical management of their access to the Internet.

References

Appadurai, A. 1990. "Disjuncture and Difference in the Global Cultural Economy." In *Global Culture: Nationalism, Globalization and Modernity*, edited by M. Featherstone. London: Sage.

Appadurai, A. 1997. *Modernity at Large: Cultural Dimensions of Globalization*. New Delhi: Oxford University Press.

Barnett, M., S. Jarvenpaa and L. Silva. 2003. "ICTs, Globalization and Local Diversity." *Communications of the Association for Information Systems* 11, available at http://cais.isworld.org/articles/11-27/default.asp?View=pdf&x=51&y=10.

Braman, S. 1990. "Trade and Information Policy." *Media, Culture & Society* 12:361–385.

Fortunati, L. 1981. *L'arcano della Riproduzione.* Venezia: Marsilio (Amer. Trans. *The Arcane of Reproduction.* New York: Autonomedia, 1995).

Gergen, K. J. 2002. "The Challenge of Absent Presence." In *Perpetual Contact: Mobile Communication, Private Talk, Public Performance*, edited by J. Katz and M. Aakhus. Cambridge: Cambridge University Press.

Gergen, K. J. 2003. "Self and Community in the New FloatingWorlds." In *Mobile Democracy. Essays on Society, Self and Politics*, edited by K. Nyiri. Vienna: Passagen.

Gray, J. 1997. *Endgames: Questions in Late Modern Political Thought.* London: Polity Press.

Jodelet, D. 1989. *Les représentations sociales.* Paris: P.U.F.

Latouche, S. 1996. *The Westernization of the World: The Significance, Scope and Limits of the Drive towards Global Uniformity.* London: Polity Press.

Longo, G. O. 2002. "Mente e Tecnologia." *Pluriverso* IV(4) and V(1):135–146.

Maldonado, T. 2005. *Memoria e Conoscenza. Sulle Sorti del Sapere Nella Prospettiva Digitale.* Milano: Feltrinelli.

Moscovici. S. 1961. *La Psychanalyse, Son Image et Son Public.* Paris: P.U.F.

Nyiri, K., ed. 2005. *A Sense of Place. The Global and the Local in Mobile Communication.* Wien: Passagen Verlag.

Rakow, L. and V. Navaro. 1993. "Remote Mothering and the Parallel Shift: Women Meet the Cellular Phone." *Critical Studies in Mass Communication* 10(2):144–157.

Sen, A. 1999. *Development as Freedom.* New York: Random House.

Strassoldo, R. 1992. "Globalism and Localism: Theoretical Reflexions and Some Evidence." In *Globalization and Territorial Identities*, edited by Z. Mlinar. Albershot: Avebury.

Taylor, P. J. 1999. *Modernities: A Geohistorical Interpretation.* Minneapolis: University of Minnesota Press.

Tomlinson, J. 1999. *Globalization and Culture.* Cambridge, UK: Polity Press.

Wallerstein, I. 1974. "The Rise and Future Demise of the World Capitalist System: Concepts for Comparative Analysis." *Comparative Studies in Society and History* 16:399–406.

Wellman, B. *et al.* 2005. "Connected Lives: The Project." In *The Networked Neighborhood*, edited by P. Purcell. Berlin: Springer.

Contents

Part 1
Science, Technology, Culture, and the Body

1 Science and the Culture of Everyday Life in the Philippines

Raul Pertierra

Introduction

Science is often depicted as a form of universal knowledge whose validity exceeds the culture within which science exists. But one can also talk about the culture of science as a system of beliefs, values, practices and structures. Science as a system generally coexists with other cultural systems within broader societal structures. The values of science never entirely supplant earlier values even when science becomes a dominant system in society. Durkheim (1912) argued that science would only be able to explain religion when science itself becomes a religion. No society has yet reached that stage but in some science, as a system of values and practices, is held in higher esteem than others.

The aim of this paper is to discuss the conditions for a culture of science in the Philippines. It looks at science as an ultimate cultural value and its practice as an example of the pursuit of cognitive excellence. The paper explores the relationship between a culture of science and the orientations and values of everyday life. My interest is to investigate cultural assumptions linking everyday life to the generation of scientific knowledge. This generation of new knowledge is based on the valorization of science. It accepts global standards and insists on contributing to the expansion of science as part of a universal civilizing process (Elias 1982). It assumes a degree of reflexivity interested in duplicating and extending the scientific accomplishments of others. A culture of science is globally oriented.

Definitions and Uses of Culture

The use of culture in recent times has undergone several significant transformations. In its most basic anthropological sense, culture is the framework for organizing the world and our position in it. It is a set of principles that locate and orient human beings within their existential realities. Some of these realities (e.g., forces of nature) are pre-given, even if culture, by giving them meaning, significantly alters their implications for human society. Because culture has to adjust to given realities, its principles are never totally consistent or uniformly applied. Lived-culture is therefore frequently ad hoc, inconsistent and accommodative. Some see culture as a post-hoc reflection of established practice rather than a set of well defined and consistent rules.

By extension, a culture of science consists of principles and practices whose aims are to explore the natural world in order to bring it under human control. Notions of the natural world and demands to control it are themselves products of culture. Using a western model, science is seen as involving theoretical and empirical knowledge dealing with the world as brute facticity, and using this knowledge both as a value in-itself (an intrinsic cultural value) as well as a tool for the control of nature for human ends. Hence, science consists of understanding the natural world as an end value as well as applying this knowledge instrumentally as forms of technology.

Science in its employment as technology is a feature of all societies even if it is more developed in some than in others. But science as a form of comprehending nature as an ultimate value is a recent feature and was best expressed in western society in the 18th and 19th centuries. It is a product of the progressive secularization of European society and associated with the growth of capitalism as well as the spread of colonization. Hence, a culture of science is derived from a broader European secular tradition that views nature (and by extension society) as ultimately amenable to human understanding and control. How well a culture of science can flourish in other (non-western) societies is an empirical question. My interest is to explore the compatibility of a culture of science within the features of everyday Philippine life. Do the features of everyday life support or hinder a culture of science?

Culture and Identity

The anthropological notion of culture sees it as a mode of life. Culture is underpinned by its basis in social structure. However, modernity and globalization are characterized by widening gaps and cleavages in peoples' modes of life. Increasingly, a diverse range of social structures obliges their members to enter into "meaningful interactions" with each other. Filipino workers manufacture goods for an affluent western market whose representations and images simultaneously confirm the Philippines' backwardness and its poverty. Overseas workers return from abroad dressed in the latest fashions and laden with commodities, often produced at low cost in the Third World, which they flaunt before their impoverished and envious kin. Overseas workers reproduce their identities by acquiring symbols of Otherness (Friedman 1990), while their poor relatives can only draw from the familiar and the local. Cultural images and values are no longer directly congruent with their corresponding social structures. This disparity and fragmentation between culture and social structure is a major feature of modernity.

Globalization draws on representations and meanings often far removed from people's ordinary experience. As the field of representations and meanings expand globally, understanding them locally becomes more difficult. While there is an excess of meaning, there is also a lack of sense. A firm "grounding" in this plethora of meaning becomes increasingly problematic despite or probably because of the expansion of structures of communication. The more there is to communicate, the less able we are to understand.

Globalization also allows for cultural complexes to be displaced beyond their original social structural sources. Hence, Filipino scientists pursue their interests even if a culture of science is poorly developed in the country. But this cultural displacement prevents the full development of scientific achievement, ensuring only duplication or minor discoveries. A common complaint of scientists is that they are not taken seriously or at best are seen as minor technicians at the behest of politicians and other power holders. Science as a major factor in the control of nature or in the improvement of production rarely enters the public discourse. Instead nature and society are perceived as being under the influence of supernatural powers open only to propitiation.

Culture as Common Understanding

Reaching a common and consensual understanding (Habermas 1987) under these conditions becomes difficult. Globality makes it increasingly problematic for subjects to reach an agreement based on a common understanding of a given situation. Notions of authenticity, normative rightness and objective validity become difficult to coordinate and sustain. Culture as a set of principles that locate and orient human beings within their existential realities becomes problematic, leading to fragmented or dislocated perspectives.

As a consequence of the above, all modern societies are multicultural not only because their members migrate from different backgrounds but also because culture is internally fragmented. Agreements based on a common understanding of a given situation occur only within a given sector often across societies. The Filipino diaspora is the best illustration of this cultural sector across societies. The new communications technologies such as mobile phones and the Internet facilitate this cultural reproduction even across global spaces. Ethnic identity coexists with other identities.

In the Philippines, a community of scientists exists but whether their members subscribe to a common theoretical-practical orientation in relation to the natural world is open to question. Their members draw their orientation both from the routines of everyday life as well as a global scientific outlook. Our research indicates that the former rather than the latter often determines the theoretical-practical orientation of Filipino scientists. Most scientists openly admit that they share popular beliefs about an animistic nature. Even agnostics say *tabi, tabi po* (please excuse my intrusion) in case they encroach on the space of unseen beings. These attitudes are so ingrained among Filipinos that they should be seen as pre-doxic. They are not seen as an aspect of a cognitive order but simply as a behavioral or relational mode. It is more an orientation towards the world rather than a view of it. Filipinos are very careful not to give offence to their superiors, whether corporeal or spiritual.

Culture and Social Interests

In the Philippines, knowledge is often considered a personal asset rather than an aspect of a public world (Pertierra 2003). Knowledge is guarded against interlopers and shared only with intimates. This attitude applies to science and other forms of public knowledge. It explains the preponderance of patrons and factions even in academic and professional associations dedicated to a common body of scholarship. This politicization of knowledge invades areas of technical expertise and prevents their autonomous development. Scientific and technical questions are settled politically.

Lacanilao (1994) has discussed how this politicization of knowledge as a scarce resource to be shared only among selected participants prevents the general development of expertise. Personal ties rather than professional competence determine the structure of opportunity for scientific development.

Higher degrees are seen mainly as an exercise in accreditation instead of as a preparation for further research. Once completed, theses are generally not published and hence do not contribute to the public world of scientific knowledge but only to a personal career advancement. Scientific competence is constrained by wider cultural interests preventing its full development. The production of knowledge is not considered a common good but is seen as a private asset to be strategically guarded.

The model of knowledge in the Philippines resembles the qualities of the esoteric rather than that of rational-empirical epistemes. The former is particularistic and its value is proportional to its restricted circulation. Knowledge is a closely guarded secret possessed by the master. Its validity is guaranteed by the master's personal integrity. This knowledge often derives from scriptural or religious sources, often through direct supernatural revelation. The notion of communicating with supernatural forces is a reality accepted by most Filipinos. Miracles, visions, dreams and other signs of the supernatural are regular aspects of everyday life. This esoteric knowledge is transmitted only to deserving informants based on moral rather than cognitive competence. The fewer people know about it, the more valuable the knowledge becomes.

In contrast, a rational-empirical episteme is differently constituted. Knowledge is valuable only to the extent that it is widely circulated. Its validity is guaranteed by its public empirical nature rather than its private integrity. This knowledge arises from sources whose access is open to all qualified inquirers. Their qualification is cognitive rather than moral. The model of esoteric knowledge not only pervades everyday life but is also encountered in universities and professional associations. For example, many scientific and professional meetings begin with a prayer asking for divine guidance.

The Pragmatics of Culture

An editorial of the Philippine Daily Inquirer (October 20th, 1999) asked:

> As a nation, then, are we forever consigned to backwardness and pre-modernism, bound to commit errors of judgment and short-sightedness because we have failed to develop a scientific attitude that can explain the world and predict its vagaries?

Only if the world is perceived in certain ways (e.g., unambiguous realities, regularity and predictability, falsifiability, disenchantment) is it likely to result in a scientific attitude. While culture is only one element that shapes this perception, its salience is undeniable. Many of the world's vagaries may be avoided by a better knowledge of its governing structures. Globalization and the new technologies, with their capacities and necessities for coordination, only exacerbate the imperative for a scientific orientation.

The process of secularization during the 18th and 19th centuries in Europe resulted in a broad agreement not to include supernatural elements as part of a rational-empirical orientation to the natural and social worlds. Spirit encounters were a regular feature of European life until the mid 18th century when they ceased to be reported as part of everyday events (Schneider 1993). Earlier, in 1618 scholars at Leiden University were granted special rights to investigate the natural and social worlds (Pertierra 1997). Eventually, these two elements resulted in the separation of the private and public spheres. The former (private) referred to beliefs and values that could not be verified in an empirico-rational mode, while the latter (public) represented knowledge and values that could be established by science within a political-legal order. Religion and morals were consigned to the private sphere while science and rationally based consensual laws determined the public sphere. The latter no longer privileged private knowledge, experience or morals. Instead, claims within the public sphere had to be validated according to the criteria of science or of democratically established laws.

Science and Global Culture

Globalization is exerting increased pressures for the Philippines to improve its technological expertise. While the rhetoric on the importance of technology is extensively circulated in the media, there is very little investment in the scientific infrastructure of the country. The government's earlier claims of making the Philippines the region's IT center by 2004 is unfulfilled. Any objective assessment of the country's scientific expertise points in the opposite direction. The index of scientific publications is one of the lowest in the region. In the period 1981–1992 the Philippines contributed 4% of the refereed science publications in Asia. Singapore with a population under 4 million people contributed 10% (Lacanilao 1994). Since then Singapore has doubled its number of publications from 502 to 1270 while the Philippines increased from 209 to 224 (Lacanilao 1999). The OECD recommends a minimal investment of 1% of GDP for R & D; in 2003 the budget of the Department of Science and Technology was 0.15% of GDP. The United Nations estimates that 4 scientists per 10,000 are needed to maintain adequate standards. The Philippines has less than 2 scientists per 10,000.

The Interest in Science

There are many reasons for the lack of interest in science in the Philippines. The relatively low salaries and social status of scientists as well as their lack of influence in public life, discourages the pursuit of scientific competence. Why is this competence given little social or cultural value? Is this because scientific knowledge depends on objective and impersonal criteria rather than on the personalized networks Filipinos normally use for success? Is the lack of a scientific orientation partly cultural? Cultivating skills in science only make sense if corresponding structures ensuring appropriate rewards exist. These reward structures are generally provided by the state or by the private sector. Neither is significantly effective in the Philippines.

The Philippine State has never shown a great interest in science (Pertierra 2003). The American colonizers tried to instill an interest but Philippine political culture and the local economy saw little need for science. Institutions of science established by the American colonizers quickly deteriorated when local officials took over. They saw no advantage in carrying out research whose benefits were not immediately obvious, preferring instead to import technology as the need arose. Even adapting this technology for local use is not developed. The passenger jeepney (American surplus jeep) is a standard example of such an adaptation but its use has long since been superceded. Attempts to make jeepneys more comfortable by providing air-conditioning have not succeeded. More recently, Filipinos have adopted the Tamaraw FX (Toyota local van) to replace the jeepney instead of designing a more original and effective replacement.

The private sector, given the undeveloped nature of the economy, has expressed even less interest in science. New areas such as IT are seen as an immediate source of profits rather than an opportunity to develop new global structures. Given the poor standards of education, the Philippines mostly graduates low level technicians rather than advanced computer engineers. The IT boom has resulted in call centers where operators take advantage of cheap international communication services. English language rather than technical skills are the main qualifications needed and only between 1 to 3% of applicants meet the required standards.

Role of the Middle Class

One of the major questions facing sociologists in the Philippines is the extent and significance of the middle class. Most theorists of modernity point out the crucial role of the middle class not only in providing society with its professional skills but also for instilling the value of professional work and the rewards of individual achievement. People born to wealth or those trapped in a cycle of poverty tend to view the world as constitutively determining their future. Both the wealthy and the poor often depend on private networks, rather than on achieved and public criteria of competence.

If life in the palace or the village is characterized by the intimacy of kin and consociates, modern middle class life consists mainly of interactions with strangers. Such strangers exchange services on the basis of formal rules and criteria of competence, while simultaneously preserving their anonymity as part of urban life. To facilitate such exchanges, members of the middle class resort to a conscious strategy of politeness. Much of middle class life consists of learning complex codes of behavior (linguistic, cultural and social) appropriate for such public interactions (e.g., how to address strangers such as customers and clients, how to express intimacy and formality in public, how to deal with superiors such as managers, administrators or government officials). Appropriate behavior in these situations differs from that in the domestic context. Learning these social skills is known as secondary socialization, involving specialized institutions removed from domestic relationships such as schools, work, prisons and other complex organizations.

The public sphere, within which interactions with contemporaries are conducted, is characterized by an elaborate set of formal rules. These rules are part of a generalized body of knowledge whose perspectives allow subjects to view the world from interchangeable positions (e.g., an official in one situation may become a client or a customer in another). In contrast, the private sphere arising out of direct consociation, assumes a fixed perspective based on personal and contingent knowledge. The formation and orientation of middle class interactions are marked by an acute awareness of applying formal and impersonal rules of behavior. Many members of the middle class even apply these formal rules in the domestic context, by emphasizing a discourse of excessive politeness (e.g., would you mind washing the dishes tonight).

The Middle Class and the Public Sphere

The middle class is a product of the public sphere and defines itself in its terms. For members of this class, the world is not constitutively given but must be created through rule-governed actions. While their members naturally have private interests, these are ideally separated from their public duties and expectations. For this reason, middle class life revolves around the distinction between domestic-private concerns and publicly appropriate behavior. Such a distinction becomes crucial in the context of urban life, where most interactions occur between strangers rather than among a palace coterie or co-villagers. Middle class families often train their children for future interactions by adopting formal language more suitable for strangers than kin (e.g., May I please leave the table; sons address their father as Sir).

In the political sphere, the middle class is supposed to instill the respect for law and for individual rights, these being the foundations of a democratic polity. Theoretically, such a polity is formed by consensus where each party formally establishes its legitimate claims through processes of argumentation. That is, through processes which involve abstract and general rules, implemented by unknown functionaries rather than private decisions reached on the basis of hereditary claims. The middle class is also responsible for ensuring the independence of the public sphere, with its guarantee of universal rights for all individuals, irrespective of birth or status. Most of these qualities of the middle class are associated with its strong emphasis on formal schooling as the determining factor for allocating social roles and for inculcating cognitive structures favoring abstract modes of thinking. The middle class encourages a culture of excellence as a way of legitimating its orientations and interests.

Culture of Excellence

Institutions such as the National Commission for Culture and the Arts (NCCA) as well as more specialized bodies like the Cultural Center of the Philippines have been entrusted with the generation and transmission of the nation's artistic heritage. Much of this heritage includes the western canon but there are increasing attempts to supplement it with indigenous experience. While their rank and file must necessarily draw on the middle class, policy making is still controlled by people drawn from the country's most privileged elite. This is illustrated by the personal networks which control appointments to senior positions in government as well as the private sector. Positions of importance are decided on the basis of political or social connections. The same may be said of the Department of Science and Technology (DOST) and other technical associations.

As Philippine society becomes more complex and globalized, there is a greater need for technical expertise. But despite their growing influence, Filipino scientists lead an economically precarious existence. Since their natural domain is the public sphere, often drawing their salaries from the state or other non-profit institutions, they are often persuaded to mute their social criticisms. The general weakness of public institutions in the Philippines, in particular those concerned with the generation or dissemination of knowledge such as science, disempowers members of the middle class.

The importance of the middle class for a culture of excellence has been mentioned but their socio-critical function is severely constrained by their economic dependency. The structural weakness of the Philippine middle class prevents it from exercising a steering function in the nation's cultural life. Instead, this steering function is dominated by the elite's understanding of high culture, a view which sees it as a form of display, a mere representation. By contrast, the middle class views culture as involving an understanding of the fundamental artifactuality of social life and hence of the limitless possibilities for its constitution. The middle class advocates social change and transformation as an essential element of modern life. For it, conceptions of the future shape the present. By contrast, the elite and the chronically poor see tradition and the past as a justification of the present. This temporal transformation defines the transition from the premodern to the modern.

A view of culture that sees it not only as negotiable but also as arising out of rationally defensible criteria involving cognitive and technical judgments, encourages its public discussion rather than simply its private accumulation or consumption. Culture is simultaneously an appreciation of the existing human condition as well as an expression of its counterfactual possibilities. It is this latter concern which members of the middle class are best able to explore. Until their members play a more significant role in its constitution, the Philippines will remain not only a bastion of an uninformed elitism but also trapped in populist misconceptions. It will not be able to generate a culture of excellence in the sciences or the arts.

Filipino Self Images

Aguilar (1997) has argued that Filipinos often see the country as a small, weak "feminine" entity easily preyed on by foreigners and therefore requiring protective measures. The Philippines is portrayed as emotionally immature, unable to seek its own destiny in a world of predators. In actual fact, the country is as big as the U.K. and with a population (84 million) larger than Germany's. This self-perception contrasts strongly with Singapore's view of itself (4 million) as a technological and economic powerhouse in the region.

Part of this perception of weakness and unimportance is the country's inability to fully participate in the achievements of modernity, of which science and high culture are the leading indices. Only in the sphere of political participation is the Philippines an exemplar of modernity for its neighbours. But even the country's accessible opportunities in education have not translated into technical competence. Instead, large numbers of teachers, nurses and other professionals regularly depart for overseas in search of better salaries and opportunities, including working as domestic helpers. In the last case, this often results in de-skilling. It also contributes to the perception that Filipinos are only suitable for menial jobs.

Aguilar argues that part of the reason for the country's low self esteem is its insistence to compare itself with the United States instead of its regional neighbors. This American orientation is predisposed to a relationship of servility, dependence and patronage. It reproduces the hierarchic structure of local society, where subjects occupy constitutively ascribed roles.

Until members of the middle class take a more active role in shaping Filipino society, culture will remain trapped in traditional hierarchies. Arrangements of the past rather than possibilities of the future will continue to determine the present. Neither a culture of excellence nor a culture of science can thrive within traditional hierarchies. The Philippines will only be able to look with envy not only at the achievements of the United States but even its local neighbors.

Natural and Cultivated Skills

The view that Filipinos are uninterested in science is prevalent and contrasts with the equally strong perception of the Filipino's musical and social skills. This self-perception indicates an imbalance in Philippine culture. It indicates that Filipinos are more interested in manipulating social relationships than in controlling the forces of nature. While finely honed social skills constitute a marked feature of everyday life, understanding natural causes receives far less emphasis. The former evokes creative responses while the latter reinforces acceptance and fatalism.

Many aspects of Philippine culture and social life are easily recognized as modern and global but the lack of interest in science points to a more traditional orientation. Moreover, these developed musical and social skills are generally seen by Filipinos as being naturally endowed rather than gradually achieved through a long process of disciplined learning. Networking and negotiating skills are necessary for everyday life. Most Filipinos learn these skills as an ordinary aspect of social life. The variations on the notions of *pakikisama* (cooperation) and *pakikipagkapua* (empathy) (Enriquez 1990) indicate a nuanced orientation towards others. This orientation includes performance skills such as singing, dancing, oratory and other abilities to entertain, amuse, praise or disparage. Although Filipinos are generally highly skilled in these performances, they are nevertheless seen as natural or ordinary accomplishments.

The lack of a conscious and developed culture of cultivation limits the achievements of Filipinos. It constrains them within the limits of ordinary convention seen as a natural skill. It encourages under-achievement and reproduces traditional hierarchies. For science and excellence to prosper, culture has to be seen as a field of cultivation requiring disciplined learning and exemplary standards. In other words, a culture of science must see culture as amenable and subject to an increasing rationalization. Science uses culture as a way of understanding the natural world in order to bring it under human control.

Culture of Disenchantment

The notion of the disenchantment of nature as a condition for the rise of modern science was one of Max Weber's (1978) main sociological interests. The progressive de-animation of nature and its replacement by abstract, universal laws was a major feature of the Enlightenment which, according to Weber, ultimately led to the rise of western science.

The interpenetration of the sacred and the secular were slowly disentangled. This eventually resulted in separate domains. The public and secular world of science dealt with empirical phenomena governed by discoverable regularities and abstract theories. This knowledge had to be publicly validated and any restrictions on its pursuit were strongly resisted.

Alongside this public, secular world was the private sphere, consisting of deeply held beliefs about a reality whose existence was no longer commonly shared. Religion, until then a public reality, was consigned, after generations of fruitless and irresolvable conflict, to this private sphere. Universities, hitherto repositories of knowledge of the sacred, shifted their attention from theological argumentation to science and public administration. Surprisingly, both spheres flourished, each enriching the other. Natural science was complemented by the flowering of a critical, aesthetic reflectivity in areas of life such as art, music, literature and religion. The rise of the social sciences was one expression of this new reflectivity which partly bridged the two separate domains (Pertierra 1997).

Local Informants

We interviewed prominent scientists in leading Philippine universities to inquire into their notions of everyday culture and its relationship to scientific achievement. Students in these institutions were also interviewed. While the sample was limited, it is sufficiently representative. Most of the scientists interviewed shared conceptions of the natural world very similar to mainstream views. Despite their advanced training and even after living abroad for many years, their cultural attitudes and orientations remain within the broad stream of Filipino life. There seems no evidence of a strong sub-culture among them that differs significantly from most Filipinos. There are, however, some strong personal differences.

A leading physicist argues that "a scientific tradition does not emerge by chance in society. It results from a conscious and coordinated effort by many sectors of society (industry, government and the academy) over periods of time ... Scientific research is very difficult to perform in the Philippines — the supporting research structure has not yet been sufficiently developed." In other words, science is the product of a developed public sphere whose members coordinate their resources for collective ends. A biologist noted that "science does not appeal to the masses. Scientists are projected in the media as geeks, with no social responsibility and relevance. Our media is teeming with pseudo-scientists, who give science a bad face."

Secondary science teachers are even more emphatic about the centrality of religious belief in their lives than their counterparts at university. "I put God in everything I do" is a typical view. They readily admit that supernatural forces intervene in nature. One teacher admitted that: "Yes, everything that happens in nature is the work of God," while another was more specific: "God controls the weather. We pray to him for good weather." But while the views above are common, other teachers are less emphatic or insistent about the centrality of divine intervention.

As might be expected, students share many of the above views regarding science and the natural world. However, there is some evidence that students are developing a more secular consciousness than their elders. This is best expressed in their attitude and acceptance of new technologies. Rather than questioning the possibly disruptive aspects of new technologies, students not only willingly accept them but are quick to adjust their everyday lives around these new opportunities. Research on the effect of mobile phones on youth identity and practice is revealing how transformative these new technologies are (Pertierra *et al.* 2002).

The Assumptions of Culture

I started this paper by asking about the relationship between aspects of everyday culture and a valuation of achievement. I argued that a culture of excellence depends on supporting structures and on orientations best provided by the middle class. These structures constitute the public sphere and involve competencies learnt through disciplined study. But the Philippine middle class is still insufficiently developed. Its members have few resources and often depend on systems of patronage whose basis is rooted in private networks. The elite still control the public sphere, where patronage rather than competence is the basis of success. Even technical questions are often decided on the basis of general criteria.

A major use of culture is to refer to practices and beliefs so common and taken-for-granted that its assumptions are rarely challenged. Hence the grounds of these practices and beliefs are rarely problematized. I use this understanding of culture when looking at aspects of the taken-for granted-world.

A major element of this taken-for-granted-world is the notion of an animated and purposeful nature regularly intervening in human culture and society. Natural events such as earthquakes, volcanic eruption and floods are often interpreted as signs of a displeased nature (Bankoff 2003). Sometimes nature is seen as an instrument of higher supernatural powers. The natural world is inhabited by spiritual entities that must be propitiated to avoid their anger. Most Filipinos conduct propitiative rituals whenever they have transgressed the domain of nature-spirits. These cultural notions are so common that most Filipinos take them for granted, including scientists.

Another example of a taken-for-granted reality is the common belief that Filipinos are naturally skilled in the performance and discursive arts. These so-called natural skills do not require further cultivation and hence prevents the attainment of exemplary achievements. Many of our informants often claim that Filipinos possess world-class skills and talents but are unable to explain why these accomplishments remain undiscovered. The contribution of Filipinos to the world is readily recognized but it lies in relatively unspecialized areas such as seamen, domestic workers and care-givers. In the arts, sciences and sports, at a global level, Filipinos are clearly underrepresented. The Philippines is a society of skilled amateurs but very few accomplished professionals. Advanced scientific knowledge cannot prosper under these conditions, leaving only the cultivation of basic technical skills.

Achievements of Culture in a Culture of Achievement

For the reasons mentioned, highly developed conscious and disciplined accomplishments leading to "products" such as art, science, law and philosophy are weakly developed in the Philippines. Most Filipinos are satisfied with quickly learnt skills expressed in much popular culture rather than the disciplined accomplishments resulting in "high culture." This attitude is also reflected in activities such as sport, where general, basic or self taught skills predominate rather than the institutional and systematic training required of a globally successful athlete. Since most activities do not presume highly developed skills, many talented Filipinos switch from a field such as entertainment to art or politics.

The Philippines has competent scientists but most of them point out that their abilities are better rewarded abroad. Local scientists and other professionals feel left out of broader societal resources. The undeveloped nature of the public sphere discourages structures like science which depend on objective rather than inter-subjective knowledge.

Invisible and Contrived Cultures

Culture may be so familiar that it is invisible or, alternatively, so accomplished as to be seen as exemplary. There are striking differences between these two understandings of culture. The first makes culture disappear within the routines of everyday life while the second extricates it from daily experience in order to focus on as yet unachieved goals. In the first usage, culture disappears as an element of the taken-for-granted-world. In the second, culture is detached from this unquestioned world and oriented to another more abstract, less immediate and still unachieved reality. When exemplary standards are achieved in the arts, sciences or sport, the results may be shocking, astonishing or awe inspiring.

Most of our informants readily admitted practicing religious rituals as part of everyday life. Catholic Filipinos make the sign of the cross almost unawares, indicating the invisible aspect of culture. But our more abstract oriented informants dismissed such rituals in favor of more counterfactual ones such as praying for world peace. In the context of everyday life such a prayer requires a strong abstraction and detachment from mundanity. Culture allows us to shift almost unperceptively from a taken-for-granted-world to a more abstract and less directly experienced reality. But only a self-critical reflection will reveal this transition. Such critical reflections constitute attitudes best exemplified in the middle class.

Structures for the Development of Science

Social, political and economic institutions play a determinative role in developing an interest and competence in science and technology. The Philippine

State is a major player and often determines their practical success. The lack of government support for science and technology (0.15% of GDP) is a constant complaint of all our informants. Expenditure in research and development as well as in education is among the lowest in the region and declining further. There is little hope for political support for science and technology to improve in the near future. As our informants claim — "science is not a major interest of government" — "there are no role models of politicians in science."

Political goals in the Philippine reflect the cultural malaise discussed in this paper. These goals range from the fantastic (e.g., IT center of Asia), the impractical and unattainable (e.g., reduction of poverty without population control) or the mediocre (providing basic facilities such as water and electricity or one book per student). They suffer from an excessive imagination ungrounded in the world or alternatively, from a lack of creative and challenging insights. These goals emanate from elites unaccustomed to realistic achievements or populist demands for basic citizenship rights.

The students we interviewed, while sharing most of the cultural orientations of their elders, showed some signs of a greater awareness of global factors. While still deeply influenced by the family, many young people are making their own decisions and see science as an opportunity to declare their independence. The youth are likely to develop a less enchanted view of nature than their elders. This is as much a consequence of the growing specialization of contemporary culture as it is a rejection of earlier views. Peer influence and the growth of youth culture accentuate more material and less enchanted features of contemporary life.

Conclusion

In this paper, I have discussed the complex relationship between a culture of science and the practices of everyday life. I have shown that various aspects of everyday culture inhibit but not prevent an orientation towards science. The view of nature as animated, the ordinary and common acceptance of miraculous events, a personalized attitude towards knowledge, seeing it as an asset to be guarded against interlopers, the overlapping of the private-public spheres and the low valorization of a specialized competence in favor of generalized social skills are the main cultural factors impeding the development of a culture of science.

There are also social factors affecting science. Among them is the lack of a political interest in funding science, the undeveloped nature of an economy requiring low-level rather than high-level technical skills and the absence of institutions specializing in research. The lack of supporting structures prevents a scientific ethos from establishing local roots. In addition, a cultural valorization of competence is weakly developed. These factors are related to the undeveloped nature of the middle class and its corresponding public sphere.

Globality has led to the proliferation of cultural representations, with their corresponding meanings. These images are often disembedded from local structures and everyday experience. This is why science exists even in societies that do not have a scientific culture. But for science to prosper, the corresponding cultural incentives have to be present. Attempts to incorporate these cultural incentives into the routines of everyday life have so far not succeeded. For this to happen Filipinos have to become more aware of the invisible elements of culture as well as establish exemplary standards of achievement.

A major paradox of globality is that while it results in the universalisation of elements of culture, it also encourages their resistance. The culture of science has come under challenge as science itself becomes increasingly indispensable in everyday life. The classical distinction between science, as a disinterested pursuit of knowledge, and technology as its instrumental application, is no longer viable. Technological need increasingly determines scientific inquiry, making possible new forms of liberation as well as fears of domination. Genetics creates a post-human subject combining the immediacy of a tool with the effectiveness of a machine. While we earlier controlled the machine, now the machine becomes us. Homo habilis is transformed into homo cybergensis. Social theorists (e.g., Kirby 1997) presage the end of the human corporeal and its replacement by the post-human cyborg. Over enthusiasts of science, such as Gray (2002:9) (after Babel) declares "We're going to be Gods, we might as well get good at it."

References

Aguilar, F. 1997. "Embodying the Nation." *Philippine Studies Journal* 45(2): 197–224.

Anderson, W. 1995. "Where Every Prospect Pleases and Only Man is Vile, Laboratory Medicine as Colonial Discourse." In *Discrepant Histories: Translocal Essays on Filipino Cultures,* edited by V. Rafael. Manila: Anvil Publishing.

Bankoff, G. 2003. *Cultures of Disaster: Society and Natural Hazard in the Philippines.* London: Routledge.

Durkheim, E. [1912], 1995. *The Elementary Forms of the Religious Life*, translated by K. Fields. New York: The Free Press.

Elias, N. 1982. *The Civilizing Process*, Vol. 1 & 2. Oxford: Basil Blackwell.

Enriquez, V. 1990. *Indigenous Psychology: A Book of Readings*. Philippines: New Horizon's Press.
Friedman, J. 1990. "Being in the World: Globalization and Localization." *Theory, Cutlure, and Society* 7:311–328.
Gray, C. 2002. *Cyborg Citizen*. New York: Routledge.
Habermas, J. 1989. *The Theory of Communicative Action*, Vols. 1 & 2, translated by T. McCarthy. Boston: Beacon Press.
Kirby, V. 1997. *Telling Flesh*. New York: Routledge.
Lacanilao, F. 1999. "Science in the University of the Philippines." *Diliman Review* 47(7):24–28.
Lacanilao, F. 1994. "Journals and Scientific Progress." *Diliman Review* 42(2):6–12.
McFerson, H. 2002. "Filipino Identity and Self Image in Historical Perspective." In *Mixed Blessing: The Impact of the American Colonial Experience on Politics and Society in the Philippines*, edited by H. McFerson. Connecticut: Greenwood Press.
Pertierra, R. 1997. *Explorations in Social Theory and Philippine Ethnography*. Quezon City: University of the Philippines Press.
Pertierra, R. *et al.* 2002. *Txt-ing Selves: Cellphones and Philippine Modernity*. Manila: De La Salle University Press.
Pertierra, R. 2003. *Science, Technology and Everyday Culture in the Philippines*. Quezon City: IPC, Ateneo de Manila University.
Philippine Daily Inquirer. 1999. Editorial, September 20.
Saloma, C. 2002. "Doing IT in the Philippines: Understanding Class in the Global and Gendered Division of Labor." Ph. D. dissertation, Bielefeld University, Germany.
Schneider, M. 1993. *Culture and Enchantment*. Chicago: University of Chicago Press.
Weber, M. 1978. *Economy and Society*, edited by G. Roth and C. Wittich. Berkeley: University of California Press.
Zialcita, F. 2002. "Spirits, Water Problems and Household Needs." In *Cogito Ergo Sum*, edited by Lee-Chua, Q. Quezon City: Ateneo de Manila University Press.

2 Technology Transfers of Chinese Universities: Is Mode 2 Sufficient for a Developing Country?

Wei Hong

Introduction

As an important source of knowledge and intellect, research universities play a key role in promoting technological innovations. Two decades ago, this role was exemplified by the publication of cutting-edge research findings. Although significant in the long-term, this contribution is indirect and uncertain. Therefore, in the past 20 years, universities have been expected to contribute to the national economy in a more direct way, i.e., by conducting more applied research and making the results available for commercial use. To encourage such a transition, the governments of many countries have adopted an innovation-oriented science policy. For example, in 1980, the U.S. government passed the Bayh-Dole Act, which allowed American universities to retain titles to inventions that were derived from government funding. Since then the number of patents issued to universities has increased exponentially. Patents filed by American universities increased from about 250 per year in the 1970s to over 3,000 per year at the end of the century (U.S. Patent and Trademark Office 2002). While a technology transfer office, which helps scientists transfer their inventions to industry, is now essential for every research university, only about 20 universities had such an office in 1980 (Colyvas *et al.* 2002). The technology transfer activities of universities have generated considerate license income, created new jobs, and promoted local economies. In 2002 alone, the public research sector in the U.S. received a total of $1.19 billion in license income and formed 450 new start-ups (Association of University Technology Managers (AUTM) 2002). Given these huge successes, many countries have adopted their versions of the Bayh-Dole Act (Organisation for Economic Co-Operation and Development (OECD) 2003).

China, which is in transition from a planned economy to a market economy, has a very different institutional context and level of economic development from those of Western countries, where most of the studies on innovation have been conducted. Parallel with the economic transition, the national innovation system is also under reform (Liu and White 2001), and a distinct set of technology transfer mechanisms have been produced (Young 1999). These rapid changes offer us a unique opportunity to study the institutionalization of the university's role in the new national innovation system, thus deepening our understanding of the economics and sociology of innovation and economic development systems.

Based on 40 semi-structured interviews conducted in China in 2004, this chapter describes how Chinese universities aggressively engage in technology transfer activities and analyzes the implications. In the following section, the rationale and history of two modes of scientific research are introduced. Section three briefly reviews the history of Chinese science policy and concludes that research oriented to practical uses has always been advocated by the Chinese government. Section four introduces the methodology applied in this study. The major forms of technology transfer from university to industry are described in detail in Section five. Sections six and seven discuss whether the extensive university-industry linkages are negatively affecting the institution of science and what the relationship should be among basic research, applied research, and development in a developing country such as China. Section eight concludes the paper.

Mode 1 or Mode 2: To What Extent Should Universities be Involved in Practically Oriented Research?

Gibbons *et al.* (1994) have perceived two modes of science.[1] Mode 1 is curiosity-driven research that does not take into consideration practical applications. Scientists have their own right to choose research topics, with no concern for commercial potential, political correctness, or even national interest. This mode, which guarantees the freedom and independence of science, is believed to be the right mode by traditional standards. A university is seen as

[1]This simple division is of interest in this paper because it suggests the different philosophies of scientists and policy makers. See Stokes' model (1997) of scientific activity for more a refined division.

an ivory tower secluded from society. However, Mode 1 does not preclude the university from making a potential contribution to the national economy and social welfare. In his famous report *Science — The Endless Frontier*, Bush (1945) attributed America's advantage in technology to its strong base in science. According to the linear model Bush proposed in his report, scientific findings from curiosity-driven research will be absorbed by applied science, and in turn used in industry. Although driven by pure curiosity, basic science will eventually generate useful products. The reason why we do not set a practically oriented question at the very beginning is that we simply do not know which direction will be promising. We have to begin with all possibilities and harvest the serendipity. Bush's report had a big impact on U.S. science policy. The research funding for basic science increased significantly after World War II.

Mode 2, on the contrary, is user-driven research with a specific goal in mind. Under this mode, scientists choose research topics that are likely to yield useful products; curiosity becomes a secondary concern. This mode does not necessarily result in downstream research, though. Ideally, questions raised by industry could help improve current theory or even suggest new directions of theoretical research.[2] The relationship between basic and applied research hence becomes interactive instead of linear, as implied by Mode 1.

As mentioned above, the policy makers of many countries are encouraging universities to get more involved in practical projects to contribute directly to economic growth. This aggressive stance has aroused debates on whether the university should deviate from its traditional role of teaching and research, or in other words, whether Mode 1 or Mode 2 should be adopted by universities.

Historically, Mode 2 had been the dominant form of conducting science before the institutionalization of modern science in the 19th century (Benner and Sandstrom 2000). Merton (1938) found that between 40 and 60% of discoveries in the 17th century were originally aimed at solving practical problems. To secure science's place in society as an independent and autonomous institution, however, scientists in the late 19th century sought to build the norm of pure science (Beesley 2003; Etzkowitz and Leydesdorff 2000), or the norm of disinterestedness later defined by Merton (1973). Thus, Mode 1 gradually gained dominance.

However, as international competition and fiscal constraints increased, Mode 1 was criticized as being too remote from social needs and contributing little to national competitiveness. This induced a transition in science policy

[2]For a detailed discussion, see Kline and Rosenberg (1986) and Brooks (1994).

from basic research to applied research, and Mode 2 has regained ascendancy. Recently, a triple-helix model (Leydesdorff 2000; Leydesdorff and Etzkowitz 1996) has been proposed, in which a tripartite interaction among universities, industry, and government is regarded as key to improving the national innovation system.

Similar to what happened in the 19th century, there is concern nowadays about the negative effects of close university-industry linkages on the institution of science. According to Bekelman *et al.* (2003), university-industry links are prevalent in biomedical research. About one-fourth of biomedical scientists receive industry funding. Even more scientists receive gifts from industry, have personal financial ties with firms, or have dual affiliations. Biomedical firms actively develop close relationships with academia, a development that has been especially evident in recent years. According to two surveys (Blumenthal *et al.* 1996; Blumenthal *et al.* 1986), the percentage of life science firms supporting university research increased from 46% in 1986 to 92% in 1996. This close collaboration can lead to conflicts of interest. Bekelman *et al.* (2003) found that industry-supported research is more likely to yield pro-industry results. The same conclusion was also reached by Friedberg *et al.* (1999) and Kjaergard and Als-Nielsen (2002).

Also, the commercial behavior of scientists could impede communication within the scientific community, which is essential for the development of science. According to Campbell *et al.* (2002) and Cook-Deegan and McCormack (2001), secretive behavior among academic scientists, especially among biologists, has increased significantly. The increases have been attributed to the increasingly prevalent commercial involvement in academic fields and the need to protect the intellectual property associated with scientific research. Similarly, Murray and Stern (2004) found that the granting of formal intellectual property rights to published work meant that scientists tended to refrain from using such work, thus reducing its potential contribution to the scientific community.

Finally, the pecuniary rewards of doing commercially oriented research could drain the pool of scientists doing pure scientific research. Feller (1990) found that patenting erodes academic commitment in research universities. Even if Mode 2 is indeed efficient at promoting national competitiveness in the short run, the negative effects can be detrimental to science in the long term.

Fortunately, competing evidence has also been found in other studies. Thursby and Thursby (2002) found that there had been little change in research focus as a result of commercialization. Rather, scientists/universities have become more aggressive about trying to commercialize existing research.

Breschi *et al.* (2004) suggested that practical research provides academic scientists with both financial support and inspiring questions, thus increasing their scientific productivity. Also, they did not find any evidence of a shift in research focus. Walsh and Hong (2003) found that scientific competition is critical to explaining the practice of secrecy among academic scientists; and that university-industry collaboration sometimes promotes communication. Other studies have suggested that Mode 2 has not yet become prevalent. Agrawal and Henderson (2002) argued that most scientists still focus on producing publications rather than on filing patents. Colyvas *et al.* (2002) also concluded from their case studies that scientists did not shift their interest to commercially oriented studies.

Since prior studies have mostly focused on OECD countries, little is known about university-industry linkages in China. As a country lagging behind in modern scientific research, what mode of scientific research has China been advocating, curiosity-driven or user-driven? How is it implemented? Has university-industry collaboration (if any) led to conflicts of interest or shifts in research focus, as has been the case in developed countries? Answers to these questions will complement our knowledge of university-industry linkages in developing countries and deepen our understanding of the relationship among basic research, applied research, and development.

The History of China's Science Policy: In Favor of Mode 2

There was a debate in the 1950s, soon after the founding of the People's Republic of China, on whether Chinese universities should conduct research (Chinese Education Ministry 1999). Following the Soviet pattern, teaching had been the most important mandate of Chinese universities since 1949. Although 78% of scientific personnel were concentrated in universities, universities received no research funding until 1962 (Chinese Education Ministry 1999). Research was the territory of research institutes overseen by the Chinese academies of science, local governments, and various industrial ministries. Among the challenges to the Soviet model, one of the strongest arguments was that conducting research in universities would help solve practical scientific problems. After universities were brought into the national research system in 1962, the Education Ministry approved 18 research units within 11 universities (Chinese Education Ministry 1999). It was also clarified that university research should be aimed at improving China's modernization and living standards, prescribing essentially a Mode 2 of scientific research.

This trend was further strengthened after the Cultural Revolution, which heavily impaired the national economy. The government encouraged universities to divert their focus from basic research to applied research and to intimately collaborate with industry. The universities were seen as a major force in promoting the recovery of the economy. In a 1982 conference to present the scientific findings of universities, 32 universities presented 500 innovations, half of which resulted from collaborations among universities, research institutes, and industry (Chinese Education Ministry 1999). In 1985, the percentages of university funding allocated to basic research, applied research, and development were 13, 60, and 27%, respectively. About one-fifth of the funding came from industry (Chinese Education Ministry 1999).

In March of 1986, four renowned scientists wrote to the central government, alerting the government to China's weakness in the high-tech area. This petition directly led to the creation of the National Hi-Tech Research and Development Program (863 program), primarily supporting research in biotechnology, astronomy, information technology, laser technology, automatic control, energy, and new materials, the seven high-tech areas perceived by the Chinese government to be the most important. At about the same time, the National Education Commission clarified that universities should coordinate their research according to national needs and their own specialties. Furthermore, universities should actively engage in solving practical problems for industry, which was perceived as having positive effects on teaching and research as well.

In 1991, a production-teaching-research coordinating office was founded in Beijing to promote intimate and stable interactions among state companies, universities, and research institutes. In the same year, the first national center of engineering, with the aim of conducting applied research for industry, was established at Zhejiang University. More national centers were subsequently approved in other universities. Local governments and companies also developed a number of similar engineering centers. In 1993, the central government published guidelines on the reform and development of higher education, in which promoting economic growth was formally specified to be a major mandate for universities, with the other two mandates being teaching and research. In 1995, the percentage of university funding coming from industry had increased to 47.5% (Chinese Education Ministry 1999).

Although Mode 2 has been persistently advocated by the government at the policy level, its implementation has undergone significant changes. In the former planned economy, the government designed its scientific plan according to national needs and then distributed these projects to research institutes or universities. Scientists just accepted whatever projects were assigned to them. The research results, if transferred, were probably through channels prescribed by the government. After China began its transition to a market economy, both academia and industry acquired incentives to collaborate with each other. Since then, various forms of technology transfer have emerged. Also, realizing that many research results were not fully appropriated, the government proposed new policies to encourage rather than designate academia-industry linkages.

Method

This study mainly uses data drawn from 40 semi-structured interviews with university scientists and technology transfer officers in China. Based on the factors of overall research intensity and economic development, I chose Beijing and Wuhan as the cities to study. Beijing, one of the most developed cities in China, is home to the best universities and high-tech companies. It also accounts for nearly one-fifth of national R&D expenditures (The Chinese Statistics Bureau 2001). Technology transfer activities in Beijing should be the most active in China. Although this outstanding character means that Beijing is not a representative case for China, it allows for the generation of the maximum possible forms of technology transfer to study. Wuhan is the capital city of Hubei Province. Since a market-oriented economy was first initiated in the southern and eastern coastal regions, the farther a city is from the coastal area, the less developed its economy. Wuhan, a city located in central China, should be representative of big cities developing at a moderate level. Also, R&D expenditures in Hubei province in 2000 were 3.48 billion yuan[3], which is very close to the average level of 2.89 billion yuan (Chinese Education Ministry 2001). Therefore, the level of technology transfer in Wuhan is expected to be representative of China. By choosing one outstanding city and one representative city, I anticipate being exposed to all available forms of technology transfer from university to industry in China.

Among the 40 semi-structured interviews conducted in the summer of 2004, 30 were with academic scientists from 4 universities, and 10 were with technology transfer officers from 5 universities. These interviews are informative as to the forms of university-industry collaboration, the extent to which professors are involved in commercial behavior, and the possible negative effects of commercial behavior on teaching and research.

[3]Yuan is the Chinese currency. One yuan is approximately equal to US$0.13.

Mechanisms of Technology Transfer from University to Industry

Based on Western literature, technology is transferred from public research to the private sector primarily through patents and licenses, publications and conferences, collaboration between universities and companies, and university start-ups.[4] This section examines the implementation of these forms of technology transfer in China.

Patenting and Licensing Via the Technology Transfer Office

The Technology Transfer Office (TTO) is an important intermediary agency connecting universities and firms. After the passage of the Bayh-Dole Act, the number of university TTOs in the U.S. increased eightfold from 1980 to 1995 (AUTM 1995). The major mandate of TTOs is to transfer university inventions to industry. In order to achieve this goal, TTOs focus on (1) effectively probing disclosures of inventions from professors, either by maintaining a close relationship with productive professors or by providing incentives (Owen-Smith and Powell 2001); (2) evaluating the commercial potential of those inventions and deciding whether they are worthy of application for a patent; and (3) marketing, including publicizing the disclosed inventions, seeking potential licensees, and helping scientists establish start-ups if needed.

Patenting and licensing are new types of commercial activities for Chinese universities. Patent laws were promulgated in China in 1984. TTOs in China also work in similar ways as their Western counterparts, although more aggressively, in the hope of catching up quickly.

TTOs in China elicit disclosures of inventions in various ways. Besides government propaganda on the importance of intellectual property, experienced TTOs give their own lectures for professors, explaining the necessity of legal protection for their intellectual creations. In one university with the best engineering school in Wuhan, one young scientist said that, "Scientists in our university didn't put much emphasis on patents before. But after Mr. Feng (a technology transfer officer) gave a series of lectures, we learned that patents are important." TTOs also provide strong economic incentives to encourage the disclosure of inventions. Among the five Chinese TTOs I visited, all of

[4] Providing trained personnel is also a key form of technology transfer from university to industry. However, it is difficult to identify this form of technology transfer and evaluate its effectiveness without asking the graduates themselves. Therefore, this chapter does not include this form of technology transfer.

them pay the patent application fees and protection fees of professors for the first three years. One of the best universities in Wuhan even gives a 5,000 yuan[5] reward for receiving an invention patent. These incentives are very effective at promoting the filing of patent applications, because they eliminate the financial burdens involved in doing so.

There are three kinds of patents in China: invention patents, utility patents, and design patents. While invention patents go through a careful review process of many years, utility patents are virtually granted upon application, although a processing period of 1–2 years is inevitable. To shorten the waiting period, inventors are therefore encouraged to apply for both an invention patent and a utility patent for one invention. One professor agreed with this strategy stating, "Because it takes so long to get an invention patent, it's better to apply for a utility patent at the same time. Then my invention could be under legal protection as soon as possible." However, another professor said, "I myself am confused about the number of patents I have, because the TTO always asks me to apply for two patents for the same invention. They just want a higher number to show off."

The efforts of the TTOs have made a difference. As shown in Figure 1, while the number of patents filed by Chinese universities was quite stable before 1998, it has increased significantly since 1999.

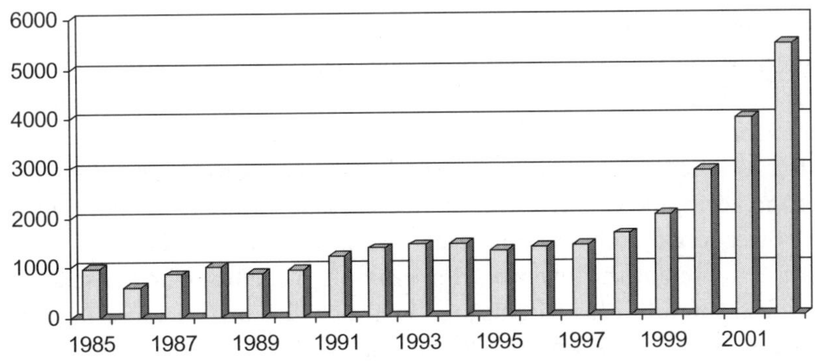

Figure 1. Patents Applied for by Chinese Universities
Source: China Patent Infonet, http://www.patent.com.cn/

[5] According to a lecturer interviewed in the same city, 5,000 yuan is about four times his monthly salary and two times an associate professor's salary. A distinguished professor said that her monthly salary is 3,500 yuan. In addition, as a full professor, she has 2,000 yuan in fringe benefits.

When asked why they applied for a patent, the responses of the scientists varied. The most frequently cited reason was protection. One scientist responded, "We now realize the importance of a patent. If I don't apply for a patent for my invention, other people might patent it and prohibit me from using it. I've heard several such stories." Another important reason is that patents have become a criterion of success in completing publicly funded projects. In China, after each publicly funded project is finished, the principle investigator will invite peers in that specific area to assess how successfully the research project had been completed. A good assessment not only contributes to promotion, but also enhances the chances to get more funding in the future. The scientific and technical significance of the main research result (*chengguo*) has always been the most important criteria in assessing the project. Now the number of patents applied for has also become an important indicator of success. A patent has also become an evaluation item contributing to promotion, just like a publication. Therefore, applying for a patent is a good way to accumulate academic capital.

Commercial gain is also a good reason for applying for a patent. One scientist who successfully transferred several of his inventions to industry said that patents generate money that he can use for further research. Another scientist I interviewed obtained five patents with his colleagues. The research group he belongs to has founded a company to commercialize a product it developed. That company has attracted two rounds of investment because of those patents. That scientist said, "Investors like patents. The marketing situation of our patented products is now really bad. But if we can have an Initial Public Offering (IPO), the number of patents we have will significantly enhance our stock value."

For American TTOs, evaluating disclosed inventions case by case is important to achieving success in licensing. In contrast, Chinese TTOs lack sufficient man power to carefully examine each invention. The market value of inventions is evaluated by the inventors. For example, a chemist who has successfully transferred several of his patents said, "As a professional, I can estimate how much my invention is worth. I'll come up with a range. Usually, I ask for an amount close to the upper range; then the potential licensee will bargain with me." After the agreement is achieved, TTOs then provide legal services, ensuring that the university obtains a fair return as the owner of the patent.

One major marketing strategy of TTOs is to build stable relationships with certain regions. Many provinces, cities, or even counties are searching for new technology for their local industries. Their government officers not only invite universities to show their latest technology, but also visit universities to seek promising inventions. As a result, technology transfer officers either travel frequently to different cities to introduce their available technologies, or arrange on-campus meetings between visitors and relevant professors. According to their needs, many cities will choose several universities as their long-term partners, helping the local firms with their R&D projects. This long-term collaboration not only facilitates the licensing of available technologies, but also guarantees stable funding from industry. One technology transfer officer said, "I have no interest in one or two small projects. What I am interested in is a long-term relationship that will bring in stable money. If we can develop one or two such regions every year, we will be hugely successful." If the American TTOs are licensing their inventions in a retail way, then the Chinese TTOs are trying to sell their technologies in a wholesale mode. However, this mode does not always work well. One officer complained that many cities have no strength in a particular industry, "They have every kind of industry, but none of them is big enough (to conduct R&D), which makes it really difficult to find a matching academic partner for them."

Colyvas *et al.* (2002) suggested that TTOs are only effective in embryonic areas and in the fields where industries do not communicate enough with academia. One technology transfer officer estimated that a half of their contracts come from professors' personal networks, and he said, "If their technologies are really promising, they will probably found a start-up to fully exploit the profit. Or they can easily find a licensee by themselves. Only when they have difficulties in commercializing their inventions, do they come to ask us for help." This quote confirms Colyvas *et al.*'s findings (2002) to some extent.

Prior research (Cohen *et al.* 1998; Rosenberg and Nelson 1994) suggests that direct transfers of university discoveries into new products or processes, while important, are a minor component of the universities' contributions to technical progress. Agrawal and Henderson (2002) also found that only a small proportion of technology is transferred through patents. Since no parallel study has been conducted in China, it is not clear how important the direct transfer is, compared to other forms of technology transfer. Yet, it is known that direct transfer itself is not very efficient. Two technology transfer officers from different universities mentioned the same official report, in which a direct transfer rate of 10% was estimated for university technology. The transfer rate for patents is even lower. As shown in Figures 2 and 3, patents are only a small part of the technologies directly transferred from universities to industry, in terms of the number of contracts and the amount of money. The major part is unpatented technology called *chengguo*, a new product or process resulting from scientific research.

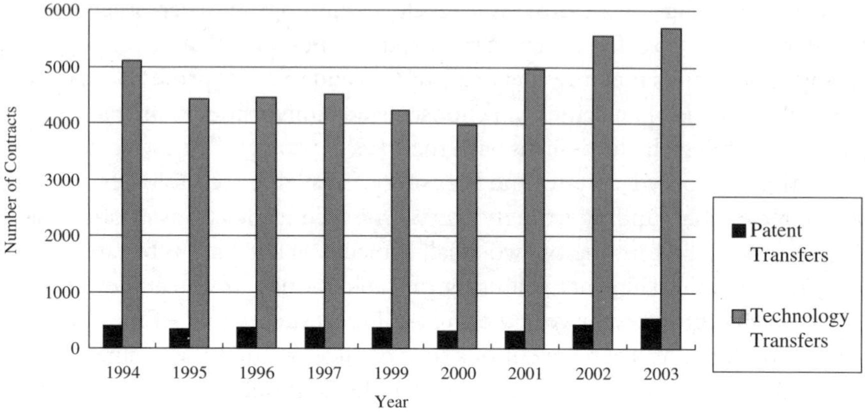

Figure 2. Number of Contracts on Patent Transfers and Total Technology Transfers for Chinese Universities
Source: University Statistics on Science and Technology 1994–2003 (1998 statistics are not available)

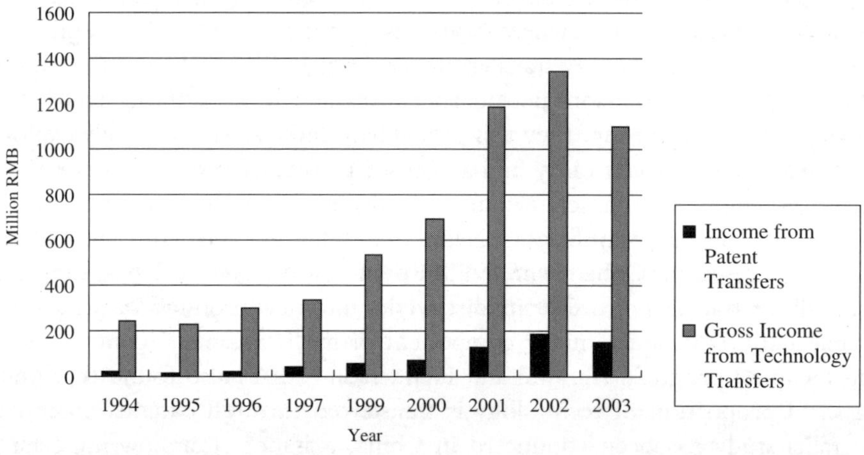

Figure 3. Income Generated from Patent Transfers and Total Technology Transfers for Chinese Uuniversities
Source: University Statistics on Science and Technology 1994–2003 (1998 statistics are not available)

Obviously, most university technology is transferred as *chengguo* rather than through the patent process. One scientist explained the difference between a *chengguo* and a patent like this:

> A *chengguo* is a complete product or process that can be actually used by industry, while patents are usually only a small part of a product or process.

> You can apply for a patent just for a minor improvement of a small component. That's not very useful for industry.

A technology transfer officer of a second-tier university in Wuhan explained why a part of *chengguo* is difficult to transfer:

> University A (a top 10 university in the same city) had sub-contracted a part of its project to us. After the project was finished, University A had a complete product to transfer to industry, but we hadn't. This was because the material part we had been working on was of no use when separated from the whole product.

The poor R&D capability of Chinese industry can partly explain why *chengguo* is preferred to patents. When firms have experience in developing their own products, those patents allowing further improvement might be attractive. For many Chinese firms lacking in-house R&D facilities, however, buying a complete product available for use is a wise choice.

Also, given the motivation of scientists in applying as described above, patents are often treated as an academic achievement or as a new requirement that needs to be fulfilled. Therefore, the scientists are not necessarily considering the commercial aspects of the patent when they make their applications. The money-free policy enables them to apply for as many patents as possible without worrying about the financial burden of doing so. Thus, the patents that have been applied for in recent years may be less commercially oriented than before. As can be seen from the figures, the number of university patents has increased in recent years, but the transfer of patents has been quite stable.

Furthermore, scientists do not necessarily apply for patents for their promising inventions. One concern is the weak protection afforded to such patents. One scientist said that, "Once you apply for a patent, every detail of your invention is disclosed. Then everyone will imitate you. We won't do that." Others do not think that they need the protection. One professor said that, "We haven't applied for a patent for our invention yet, because our products are difficult to imitate. I think only those with imitable products need to apply for patents to obtain legal protection."

Publications

Through a survey of the industry sector, Mansfield (1991) concluded that about one-tenth of the innovations in seven fields relied on the latest academic research. Also drawing on data from a survey of industrial R&D labs, Cohen *et al.* (2002) found that it was publications, conferences, and direct communications that contributed most to industry R&D. However, the validity of these claims in China is questionable. As Cohen and Levinthal (1990) have argued, firms need to invest in their absorptive capacity to make use of the seemingly free information. For many Chinese firms with no in-house R&D facilities, academic papers are too advanced to be used.

Nonetheless, university start-ups that should have a high absorptive capacity also do not seem to appreciate academic publications. Eight of the professors I interviewed had experience working in university start-ups. When asked to evaluate the Chinese publications in their area from the perspective of firms, a common comment was that only a few academic papers are useful for scientists working in industry. Their ideas mainly come from competitors, suppliers, consumers, and international journals.

One reason why Chinese academic papers do not contribute to industry as much as American papers do is because of the poor quality of the former. On the one hand, there is a huge demand for academic journals in China. Graduate students must have published papers to get their degrees. Professors need publications to be promoted. On the other hand, many Chinese academic journals have not established an effective peer review system. Accordingly, poor journals survive because they receive plenty of submissions; and papers of poor quality get published because there are so many low-level journals there. One scientist said that, "China produces hundreds of thousands of papers every year. Most of them are useless, written to fulfill the requirement of graduation or for promotion. Only a few of them are valuable."

Another reason could be the gap between theoretical knowledge and the contextual knowledge actually used in production. One scientist majoring in development complained that academic journals use theoretical significance as the only criteria for judgment:

> Many people working on practical projects have really good stuff. But those things could not be published because they are not theoretically significant. And those guys are often too busy to make their findings look theoretical. Ironically, those people who have done nothing meaningful get papers published in good journals. It's not because they necessarily have good theoretical findings, though. They just use a lot of jargon to confuse people.

Several other scientists confirmed that industry-related work rarely produces publishable results, because such work is often downstream work with no theoretical implications.

Interestingly, in the best university (which is also one of the best universities in China) I visited, opposite answers were given. One scientist working on sun energy said:

> Practical work doesn't necessarily mean research of a lower level. Indeed, many industry contracts only involve downstream work, such as designing a program interface, solving a problem in production. It's difficult to get those works published. But our work, though practically oriented, is theoretically advanced. It is, of course, publishable.

One fellow of the Chinese Academy of Engineering said:

> My work is about iron foundries, which is obviously industry-related. But that didn't prevent me from publishing good papers. I have been invited to present my paper in international conferences. The editor of one famous journal also invited me to contribute to a special issue at the journal's 50th anniversary.

One scientist who has founded his own start-up often assigned practically oriented projects to his graduate students as theses or dissertation topics. It turned out that several of them were evaluated as excellent theses or dissertations, because "their results were proven to be applicable by the industrial standard, which is much stricter than any academic criteria."

Finally, some papers are not useful because the key parts are kept secret from the public. Several interviewees admitted that they would not reveal details of their findings, especially when those findings have commercial potential. They will just explain the theoretical principle and give a simulated result to avoid revealing the key parameters.

University-Industry Collaboration

University-Industry collaboration has many forms. First, academia and industry can be brought together by national funds. The 863 program is one such fund, requiring scientists to apply with an industry partner to guarantee that the technology developed will be put into actual production. Sometimes firms are the main applicants; they then find their academic partners if necessary. For a small project, the pairing of one university and one firm is sufficient. For a huge project perceived to be significant by the government, competition is introduced to ensure a product of high quality. One professor majoring in electric motors received funding from the 863 program; she explained her position in the whole project in the following way:

> One automobile company got funds to develop an electrical vehicle. It subcontracted the motor part to myself and two companies. All three of us did the same thing the first year, and our results were then compared. Because

> my result was the best, I got the contract for the second year. Although the second company lost the contract, it was allowed to attend the competition for the second round. So I still need to work hard to retain the contract. I myself also chose three companies to make the controller for my motor. According to my investigation, company A is the best. But they did not seem to make much effort on that, so they lost the contract after the first round of the competition. They regretted that and are now working with me without pay. Finally, the one producing the best product with a reasonable price will win. And the automobile company mentioned above also needs to compete with other automobile companies that are developing the same type of electrical vehicle. Only the best vehicle will finally be commercialized.

Therefore, the 863 program makes companies work closely with universities. Without working with the universities, companies are incapable of developing the vehicle on their own. Without collaborating with the companies, a university can only produce a prototype, which demands tremendous effort for further development. By making them work together, the 863 program not only creates multiple university-industry consortia yielding high-tech products ready for commercial use, but also improves the competitiveness of state companies. After operating under a planned economy for a long time, many state companies have gotten used to waiting for assignments from their direct supervisor and receiving orders without having to compete. They have no incentive to improve the quality of their product (e.g., company A in the above story). The unusually intense competition will help correct their inertia and force them to face the challenges inherent in a market economy.

Nonetheless, because the government plays an important role in the allocation of funds, problems emerge when the government interferes in an inappropriate way. One professor mentioned that the government once assigned an electric vehicle project to an automobile company, but that company was interested in hybrid electric vehicles rather than in purely electric vehicles. That company sub-contracted the project to a university and then concentrated on the project they were really interested in. That professor said:

> If a company has a genuine interest in a project, it will invest a lot of money and energy. Another automobile company that is interested in developing an electric vehicle bought a battery factory simply for that purpose. Companies are committed to their own projects, rather than projects allotted by the government.

Sometimes the government prescribes an unreasonable deadline for big projects, thereby reducing essential experimental time. The above professor said:

> Many projects are rushed for a product. So for the automobile project, we had to come out with a real vehicle before conducting enough experiments, either on individual components separately or on a vehicle as a whole. Well, the project finished soon. We had a celebration party, we passed any test we need to pass, and things were ready for commercialization. The government cannot wait to announce that now we have developed a new vehicle. But, after all these ceremonies, the quality of the vehicle was found to be not good enough for real production.

Another professor agreed stating, "When everyone just finishes 95% of the job, the 5% unfinished part keeps accumulating. If that is the situation, then the final product won't be good." The first professor commented that this is due to governmental interference:

> Yet, firms do not work like this. They won't proceed to the second phase without finishing jobs for the first stage. They'll solve every problem that occurs. I feel the government should not interfere so much. Its role is, for example, to guide firms to find their academic partners. That's it. Don't oversee everything.

Second, many companies have started establishing research centers in universities to ensure long-term collaboration. One professor involved in two research centers described how research centers work:

> Take research center A as an example. It is led by both professors and industry people. We have a committee meeting every year to debrief what job we have done last year, and to discuss what projects could be good for the company in the near future. All of the projects approved will be contracted only to our university, and the company will donate an extra 80,000 yuan to the university annually. As a scientist affiliated with the research center, I often need to think from the company's perspective. Another research center is similar. We've proposed several approaches to update their equipment and production skills. We also have a seminar every year, and that seminar is open to other companies. Now companies are smart, they share information with each other to gain new ideas.

The higher the prestige of a university, the choosier the university is in selecting long-term industry partners. The best university mentioned above requires a three-year fund of 5 million yuan to establish a research center. One professor with such experience told me:

> Professors at our university are now prohibited from founding our own companies. So establishing an industry research center has become a good way

> to transfer technology from the university to industry. It's usually initiated by companies. They tell us what kind of technology they are using, and what kind of problems they are having. Then we tell them what our strengths are, whether we are capable of solving their problems. If we are satisfied with each other, they need to contribute 5 million yuan in the first three years to set up a research center. One-sixth to one-fifth of the money is overhead costs, the remaining part is contract money only for specific projects. We develop what they want, and they commercialize it. Now there are more than 30 such research centers at our university. Most of them are running great, but a few have failed. Some companies have unanticipated financial problems and can't afford the money anymore, then the research center will be terminated. Sometimes companies don't get the product as expected, then they'll quit.

In summary, research centers consistently provide companies with theoretical guidance and technical assistance. Through collaboration, companies not only have actual products developed with university resources, but also substantially enhance their own developing capabilities by sending employees to university labs to receive training. This type of long-term collaboration also reduces opportunism or communication difficulties often found in one-time collaborations.

A few big companies have also set up funding arrangements in prestigious universities. They regularly review proposals from scientists and fund the most interesting ones. This kind of funding provides financial assistance for theoretical research with potential for application.

Third, taking on a contract project from industry is quite common for Chinese professors. Many firms do not see a long-term relationship with a university as necessary. They seek help from a university only when specific technical difficulties appear. Thus, firms have their problems solved at a relatively low cost, and professors get research funding to support their labs. For most professors who lack the reputation and capability to attract major industry sponsors, the money from an individual project is important to maintain a research group. However, once professors have other sources of funding, these projects will probably be given the least priority. One professor claimed that she would not accept any small projects from industry because:

> The industry people are very demanding. They keep raising new requirements. It's so tiring working with them. And they can't wait to make money out of everything we just developed. But the amount of money they provide is small, usually between 100,000 and 200,000 yuan. Since I've gotten enough funding from the 863 program, I don't want to accept contract projects anymore. I'd consider it if it's big enough, though; say, above one million yuan.

Another reason why professors do not like individual contract projects is their downstream nature. One professor recalled that he had been working on small industry projects for years before he finally got a stable industry partner:

> At that time I put more effort into applied research and development, because I had no money. I had to work on projects contracted by firms to get some money. Now my funding condition is much better, so I want to shift my focus to basic research. I'm getting old; I want to do some good research.

However, contract projects can also be theoretically advanced. Several interviewees admitted that some problems raised by industry were very hard to solve. Those firms even organized a public bid to decide to whom their projects should be contracted. Therefore, university-industry collaboration in the form of a contract project is not necessarily of lower quality. A scientist commented that she would keep working on industry projects even if she had other funding, but she will screen for projects that fit her major research interests and involve advanced technology.

Besides, contract projects could be the beginning of a long-term collaboration. One scientist said that, "Our collaboration with SIEMENS was initiated by a contract project. Since they were satisfied with our progress in the first contract period, they renewed our contract and we've been keeping this relationship for eight years."

The dark side of this project-based collaboration is that both parties could be opportunistic because further collaboration is often unanticipated. It is quite common for a firm not to pay the full amount of money listed on its development contract with universities. A technology transfer officer said that:

> It has been a big problem. Now we require that firms must pay 40–50% of the contract money at the beginning. We will ask professors if that amount of money is okay with them, because they'd better not be expecting the remaining money. But many professors don't understand that. They really need those projects and they tend to trust those firms, so they insist on signing with a lower payment. Then what can I do? I can't hold the contract if those professors want it to be approved. But if they regret that later, we won't help them to sue those firms. Lawsuits are too time consuming. That's why we emphasize the first payment; at least we don't lose too much if cheating happens.

A scientist responded that one firm disappeared without paying the total amount of money, and she did not bother to search for it. She said, "I'd rather use that time to do other things more meaningful." This sort of unethical behavior is mostly attributed to small private firms.

A professor also reflected on the opportunism of scientists:

> Some professors promise everything to the industry people when they have business meetings with them. However, they cannot always deliver on those promises. One of my friends did that once, but then the company didn't come back. Later I heard that the company had figured out from their conversation that my friend was not working in that area.

Another scientist made a more severe criticism:

> Some scientists talk boastfully. A number of companies went out of business because of them. The firms concluded that extreme care should be taken when using technology developed by a university. Many companies have this consensus. This is the scientists' fault.

Exaggeration of university technology is only one reason contributing to failures in technology transfers. Another important reason is the reluctance of professors to conduct further development for firms. To apply laboratory technology in actual production, tremendous effort also needs to be devoted to medium-scale and large-scale experimentation because many parameters in large-scale production are different from those in laboratory settings. In many circumstances, firms need the help of inventors for further development. Nonetheless, professors often do not want to spend so much time on technologies that have already been sold unless an extra bonus is provided (Jensen and Thursby 2001). This also conveys to firms a negative image of professors. Indeed, all scientists with successful experiences in technology transfers have devoted a lot of time and energy on the further development of their laboratory technologies. This commitment makes a long-term relationship possible.

University Start-ups

Most university start-ups originated as research labs developing technologies with commercial potential. In their initial stage, start-ups often remain on campus, sharing resources with the research labs. Actually, in many circumstances, start-ups and research labs are two sides of the same coin. They have the same people working in the same office, fulfilling the roles of professor and company employee/employer at the same time. These start-ups are necessarily small, receiving limited orders for their products.

One scientist working in this type of start-up reported that:

> Our lab is a research setting within the university, focusing on rapid prototyping. But we also produce and sell rapid prototyping equipment, like

> a firm. Mr. Han, the head of our group, has strongly proposed combining research and production. He set up this firm to promote our research products, and we can use the money to maintain this research group. This is a big group. We have over 100 graduate students, over 10 professors, more than 10 engineers, and around 20 factory workers. We also hire dozens of retired professors. We design all of the equipment to be sold. For those parts involving little technology, such as the shell and chassis, we draw a design map and contract them to others. The key parts, such as the laser machine, are bought externally. Then we combine all the components and debug it.

This company only sells several dozens of complete units every year. Still, since the price for such equipment is high, the firm has annual sales of over 10 million yuan, which is enough to support the group. When demand increases to a certain extent, however, it becomes difficult to maintain the dual entity. One scientist selling software said:

> When I visited The Ohio State University in 1996, an American professor told me that it's unrealistic to develop software for commercial use within campus; a firm must be set up for that purpose. I didn't feel that way at that time. But with the quick pace in the development and updating of software, I feel the need to set up a separate firm taking care of these things. You know, at the beginning, we only sold three to four software packages per year. That's not difficult to maintain and update. But now, our clients keep increasing, and we need to update our software more often. I still need to teach and do research ... We just decided this year that we'll search for a young associate professor to set up a firm. We'll still do the coding job, but marketing and maintenance will be the firm's job.

In addition, in response to the over-involvement of professors in commercial activities, many universities are trying to separate start-ups from labs. Most universities require their start-ups to move off campus. This is only a physical separation, however. Many professors have moved out with their firms, but they are still affiliated with the university. These professors spend most of their time conducting R&D for start-ups, but they also advise graduate students and occasionally teach a course. Therefore, they still have a dual identity to some extent. However, with the increasingly stringent system of evaluation in universities, professors must be more productive to keep their academic positions. This is making it more and more difficult for them to play dual roles. A few universities have imposed a real separation of roles. Their start-ups must completely separate their properties and personnel from the university. The professors then must choose whether to stay on as a professor or to go out to work in the start-up.

Quite a number of start-ups that have moved out have grown substantially. There were 39 university start-ups listed as public offerings (Huang 2002) in 2002. A vice general manager of one publicly listed start-up introduced how they went through this process:

> Our company is actually composed of four start-ups. Three of them were initiated by the heads of their research groups to commercialize their research results. They themselves had strong incentives. Two of them began with left-over research funding, and they then earned some money little by little. The fourth one gained financial support from the university directly. Because it was a large national project and the research result was found to have commercial potential, the government wanted to make the technology available to the whole society and was willing to found it. The four start-ups had been doing pretty well before we had our IPO. But, as university start-ups, they were somewhat different from other firms in that their managers were also professors. They were satisfied with their situation and didn't want to take those risks to enlarge the production scale. So their firms were stable, but small. Since we had our IPO, we have gained over 400 million yuan, which enabled us to invest in our production facility. Using that money, our production base has increased from 10,000 square meters to 60,000 square meters. Our production value and sales have increased from 100 million to 400 million yuan, and our production capacity has increased from 100 million to 700 million yuan.

The Impact of Commercial Activities on Academia

As shown in the previous section, Chinese universities are maintaining a close and extensive relationship with industry. One professor said that most professors have some industry ties. The concerns of many people that money is eroding the commitment of some scientists to research are well-founded. One young scientist commented, "Everybody wants to earn money from their research. I won't stay here if I can't earn enough money." A scientist working in a top 10 university said that:

> If you are really good at research and you can get enough research funding, you don't need to do industry projects. But most professors at our university need to attract industry funding, otherwise we have no money. You know, even if you get national funding, you can't secure it every year. So you must consider other funding sources. The problem is that many professors now only focus on development; they no longer care about scientific research. I'd say 80% of professors at our university are like this.

Another retired scientist complained that earning money from the company he founded was a slow process. He said, "It's faster to earn money through doing research." This is because he could use the money left over from his research funding. He represents the scientists who see doing research as another type of business.

Business concerns also seem to impede the sharing of information. Several professors reported that companies review their papers before submission. As mentioned before, some authors do not provide real data in their papers. Many people also hide their design maps from colleagues, worrying that other people will steal their designs with commercial value. One old scientist lamented, "In the past, I organized a discussion session every two weeks; thus, we were able to share our experiences. But now, I'm not able to do it, because nobody wants to talk."

However, such secretive behavior is more likely to be characteristic of scientists working on downstream technologies that are easy to imitate. For advanced technologies or technology requiring multidisciplinary collaboration, communication is enhanced rather than constrained. One professor working on Chinese herbal medicine, a hot area with promising commercial prospects, said:

> Modern research on Chinese herbal medicine is intimately related to molecular biology. It enables us to do experiments at the molecular level, which reduces experimental costs and enhances the quality of our work. So I'd like to collaborate with biologists. Now I have ties with the biology department of Wuhan University and Beijing Normal University, and the medicine school of Tianjin University and Huadong University of Science and Technology. I also have contacts with Plymouth University in Britain and NIH in America. I just recommended one of my students to pursue a Ph.D. in the biology department of Wuhan University. The biology people also want to do some research on Chinese herbal medicine. My student should be a bridge between us. I have also sent my students to Tianjin ... Having contacts is a good thing; it broadens my imagination. By observing what other people are doing, I can get ideas about my own direction.

Another renowned scientist closely working with industry said:

> We have been to the best optical lab in Germany. We weren't secretive about our research, and neither were they. This is because we both know that, when research is up to this level, no one can steal. We both have our own strengths, which cannot be learned in a short time. So we both share the best part with each other.

Mixed evidence was also found on whether scientists shifted their research focus to a promising area. As mentioned above, one scientist claimed that the majority of his colleagues had shifted their focus to downstream research projects from industry. The group he belongs to also abandoned its 10-year experience in laser processing and shifted to the hot area of nanomaterials. By contrast, one scientist said, "Since I have decided to stay in university, academic achievement is of course important for me. I won't accept industry projects just for money; they should at least be relevant to my research. If I only want money, I would find a job in industry." The unpredictable future of a technology also prevents scientists from shifting. Scientists now enjoying their success did not necessarily expect economic returns at the beginning. Also, sometimes shifting is simply not feasible because one lacks the relevant background knowledge, experience, and perhaps, equipment.

When asked about the impact of commercial activities on teaching, many professors responded that positive and negative effects coexist. On the one hand, in Chinese universities, the teaching materials for a course are often outdated. Therefore, commercial projects allow professors to teach new things from their actual experience. On the other hand, commercial activities reduce the time professors spend on work related to teaching. One professor estimated that one-third of his time was spent traveling; thus, he had insufficient time to prepare for class. Based on his extensive industrial experience, he could have written a better textbook in his area. But time constraints made this impossible.

Interaction between Basic Research, Applied Research, and Development

The extensive collaboration between Chinese universities and industry confirm that a Mode 2 of scientific research has been adopted in China. University scientists are actively responding to questions raised by industry. This section discusses the implications of a Mode 2-oriented science policy.

As a result of this Mode 2-oriented policy, most resources in Chinese universities are allocated to applied research and development. On the one hand, the interaction between applied research and development has been strengthened, as reflected by the active technology transfer activities previously described. Many engineering majors, such as those in mechanical engineering, have their own production factory for small-scale production. While these factories were traditionally used as production bases for students to acquire hands-on experience, they now enable research laboratories to directly transfer their technologies to products for sale. That is why many Chinese professors in engineering schools do not see their research as being very different from their industry projects. They regard their commercial activity as a good way to validate their research, and this interaction is certainly beneficial for the improvement of technology. In this sense, Chinese universities successfully combine their applied research with industry development and benefit from university-industry links.

On the other hand, the link between basic science and applied science in China is relatively weak. Because of the poor funding situation, basic research in China has little effect on applied research. This could partly explain why Chinese industry, even with the help of scientists, relies heavily on reverse engineering and importing to improve technological levels. Without a strong foundation of basic research, applied research and development lacks the stimulus essential for innovation.

Because China is still a developing country, Chinese scientists agree that applied research should be given higher priority. A physicist commented:

> Although I'm doing basic research, I agree that our country should devote more resources to applied research. We have such a huge population, and many people are still very poor. We should focus on applied research and development to solve the most urgent problems. As for basic research, it's quite open. With access to international journals, it's easy to duplicate those results. We don't have to spend a lot of money on it. We just need to maintain a basic research team for educational purposes.

Scientists working on applied research judge research on the basis of its practical value, with many of them claiming that "research yielding no usable technologies is meaningless." Industry scientists show more distance from basic research. When they encounter technical difficulties related to basic science, they often choose to buy a module from abroad to avoid working on the tough part. In contrast to the intimate relationship between applied research and development, basic research is somewhat isolated.

Although it is understandable that China is devoting its limited resources to practically oriented research, the weakness in basic research could impair China's long-term competitiveness. Both the government and the scientists are taking a free-ride strategy, relying on the international scientific community to conduct basic research. However, at least two factors invalidate this strategy.

First, various studies have proven that knowledge spillovers are localized, even though the knowledge can be accessed anywhere (Adams and Jaffe 1996; Audretsch and Feldman 1996; Feldman 1994; Hicks, Breitzman *et al.* 2001; Jaffe 1989; Jaffe *et al.* 1993). By using patent bibliometrics, Jaffe *et al.* (1993) found that the impact of innovations is limited to a local area, and that spatial diffusion takes time. Hicks *et al.* (2001) showed that this is not only true for technology, which might include tacit knowledge necessitating geographic proximity, but is also valid for basic research. They found that corporate patents cite more locally produced academic papers. It is therefore doubtful that China could benefit as much from basic research conducted abroad as those countries committed to basic research, although the research results are accessible and inexpensive.

Second, Cohen and Levinthal (1990) have argued that absorptive capacity is important in innovations. Although information seems to be free and abundant, firms must invest in their knowledge base to absorb useful knowledge. Extending this concept to the national level, a country needs to maintain its basic research at a certain level to make the utilization of advanced research results possible.

Concluding Remarks

This chapter extensively describes various forms of technology transfer from Chinese universities to industry. First, technology transfer officers in Chinese universities are successful in eliciting disclosures of inventions, and they are aggressive at seeking potential licensees and industry partners region by region. Unlike the U.S., patents only account for a small proportion of technologies directly transferred through Chinese TTOs, perhaps because many professors do not consider the commercial prospects of an invention when they applied for a patent. Second, the transfer of technology via publications seems to be negligible. Academic papers are not very useful for industry because of their generally poor quality, theory-oriented nature, and concealment of key parameters. Third, university-industry collaborations are prevalent in Chinese universities. The Chinese government encourages such collaborations by making certain funds available only to university-industry co-applicants. Industry is also actively seeking R&D services from universities, inducing many long-term or short-term collaborations. Among short-term collaborations, however, opportunism constitutes a challenging problem for both firms and universities. Fourth, a number of university start-ups have gradually developed from small laboratory workshops to mature firms. Some of them have attracted investments and have been able to expand their production scale.

Although the prevalence of commercial activities does not necessarily erode academic commitment, there is evidence that some professors have shifted their research focus to a profitable direction and that economic concerns have prevented scientists from sharing information. However, these negative effects are more likely to be found among scientists conducting low-level research. For those working on upstream research projects, communication is enhanced rather than constrained.

Ideally, Mode 2 necessitates interactions between basic research, applied research, and development. This ideal type works in Western countries because they have a strong foundation in basic research accumulated from the time when Mode 1 was dominant. In addition, although advocating Mode 2, these countries still maintain a high level of funding for basic science. In China, because a Mode 2-oriented policy has been imposed since the 1950s, the basic research base has been relatively weak. In such circumstances, while the interaction between applied research and development is perhaps stronger than that in many countries, the relationship between basic and applied research is weaker, not to mention the relationship between basic research and development. Although basic research is not confined by a country's borders, the transfer of knowledge is. Moreover, underinvestment in basic science will lead to a low absorptive capacity, which will also preclude China from sharing in the most advanced scientific knowledge.

In a developing country such as China, a Mode 2-oriented science policy might be a practical choice. Nevertheless, the weakness in basic science means that China will have to follow the high-tech innovations of developed countries, which will ultimately impair China's international competitiveness. This is a dilemma for China, and for other developing countries.

This research was supported by the Provost Award from the Graduate College at the University of Illinois at Chicago. I wish to thank John Walsh, who provided guidance with the fieldwork and commented on an early draft of this chapter. I am also grateful for the editorial assistance offered by John Hessian.

References

Adams, J. D. and A. B. Jaffe. 1996. "Bounding Effects of R&D: An Investigation Using Matched Establishment-Firm Data." *Rand Journal of Economics* 27:700–721.

Agrawal, A. and R. Henderson. 2002. "Putting Patents in Context: Exploring Knowledge Transfer from MIT." *Management Science* 48:44–60.

Association of University Technology Managers (AUTM). 1995. "AUTM Licensing Survey." Norwalk, CT: AUTM.

AUTM. 2002. *AUTM Licensing Survey*: Norwalk, CT: AUTM.

Audretsch, D. B. and M. P. Feldman. 1996. "R&D Spillovers and the Geography of Innovation and Production." *American Economic Review* 86:630–640.

Beesley, L. G. A. 2003. "Science Policy in Changing Times: Are Governments Poised to Talk Full Advantage of an Institution in Transition?" *Research Policy* 32: 1519–1531.

Bekelman, J. E., Y. Li and C. P. Gross. 2003. "Scope and Impact of Financial Conflicts of Interest in Biomedical Research." *Journal of the American Medical Association* 289:454–465.

Benner, M. and U. Sandstrom. 2000. "Institutionalizing the Triple Helix: Research Funding and Norms in the Academic System." *Research Policy* 29:291–301.

Blumenthal, D., N. Causino, E. Campbell and K. S. Louis. 1996. "Relationships between Academic Institutions and Industry in the Life Sciences: An Industry Survey. " *New England Journal of Medicine* 334:368–373.

Blumenthal, D., M. Gluck, K. S. Louis and D. Wise. 1986. "Industrial Support of University Research in Biotechnology." *Science* 231:242–246.

Breschi, S., F. Lissoni and F. Montobbio. 2004. "Open Science and University Patenting: A Bibliometric Analysis of the Italian Case." In *10th International J. A. Schumpeter Society Conference: Innovation, Industrial Dynamics and Structural Transformation Schumpeterian Legacies.* Milan.

Brooks, H. 1994. "The Relationship between Science and Technology." *Research Policy* 23:477–486.

Bush, V. 1945. "Science, the Endless Frontier: A Report to the President on a Program for Postwar Scientific Research." Washington, D. C.: U.S. Government Printing Office.

Campbell, E. G., B. R. Clarridge, M. Gokhale, L. Birenbaum, S. Hilgartner, N. A. Holtzman and D. Blumenthal. 2002. "Data Withholding in Academic Genetics." *JAMA* 287:473–480.

Chinese Education Ministry. 1999. *The 50-year Development of Science and Technology in Chinese Universities.* Beijing, China: The Higher Education Press.

Chinese Education Ministry. 2001. "Statistics on Higher Education for the Past 50 Years."

Cohen, W. M. and D. A. Levinthal. 1990. "Absorptive Capacity: A New Perspective on Learning and Innovation." *Administrative Science Quarterly* 35:128–152.

Cohen, W. M., R. Florida, L. Randazzese and J. P. Walsh. 1998. "Industry and the Academy: Uneasy Partners in the Cause of Technological Advance." In *Challenges to Research Universities*, edited by R. G. Noll. Washington, D.C.: Brookings Institution Press.

Cohen, W. M., R. R. Nelson and J. P. Walsh. 2002. "Links and Impacts: The Influence of Public Research on Industrial R&D." *Management Science* 48:1–23.

Colyvas, J., M. Crow, A. Gelijns, R. Mazzoleni, R. R. Nelson, N. Rosenberg and B. N. Sampat, 2002. "How do University Inventions Get into Practice?" *Management Science* 48:61–72.

Cook-Deegan, R. Mullan and S. J. McCormack. 2001. "Patents, Secrecy and DNA." *Science* 293:217.

Etzkowitz, H. and L. Leydesdorff. 2000. "The Dynamics of Innovation: From National Systems and Mode 2 to a Triple Helix of University-Industry-Government Relations." *Research Policy* 29:109–123.

Feldman, M. P. 1994. *The Geography of Innovation*. Dordrecht: Kluwer Academic Publishers.

Feller, I. 1990. "University as Engines of R&D Based Economic Growth: They Think They Can." *Research Policy* 19:335–348.

Friedberg, M., B. Saffron, T. J. Stinson, W. Nelson and C. L. Bennett. 1999. "Evaluation of Conflict of Interest in Economic Analyses of New Drugs Used in Oncology." *Journal of the American Medical Association* 282:1453–1457.

Gibbons, M., C. Limoges, H. Nowotney, S. Schwartzman, P. Scott and M. Trow. 1994. *The New Production of Knowledge*. London: Sage.

Hicks, D., T. Breitzman, D. Olivastro and K. Hamilton. 2001. "The Changing Composition of Innovative Activity in the US — A Portrait Based on Patent Analysis." *Research Policy* 30:681–703.

Huang, W. 2002. "Reforming University Start-ups: Interview with Dr. Zhimin Li." China Education and Research Network. http://www.edu.cn/20021226/3074885.shtml.

Jaffe, A. B. 1989. "Real Effects of Academic Research." *American Economic Review* 79:957–970.

Jaffe, A. B., M. Trajtenberg and R. Henderson. 1993. "Geographic Localization of Knowledge Spillovers as Evidenced by Patent Citations." *The Quarterly Journal of Economics* 108:577–598.

Jensen, R. and M. Thursby. 2001. "Proofs and Prototypes for Sale: The Licensing of University Inventions." *American Economic Review* 91:240–259.

Kjaergard, L. L. and B. Als-Nielsen. 2002. "Association between Competing Interests and Authors' Conclusions." *British Medical Journal* 325:249–252.

Kline, S. J. and N. Rosenberg. 1986. "An Overview of Innovation." In *The Positive Sum Strategy: Harnessing Technology for Economic Growth*, edited by R. Landau and N. Rosenberg. Washington, D.C.: National Academy Press.

Leydesdorff, L. 2000. "The Triple Helix: An Evolutionary Model of Innovations." *Research Policy* 29:243–255.

Leydesdorff, L. and H. Etzkowitz. 1996. "Emergence of a Triple Helix of University-Industry-Government Relations." *Science and Public Policy* 23:279–286.

Liu, X. and S. White. 2001. "Comparing Innovation Systems: A Framework and Application to China's Transitional Context." *Research Policy* 30:1091–1114.

Mansfield, E. 1991. "Academic Research and Industrial Innovation." *Research Policy* 20:1–12.

Merton, R. K. 1938. *Science, Technology and Society in Seventeenth Century England*. Burges: Sainte Catherine Press.

Merton, R. K. 1973. *The Sociology of Science*. Chicago: The University of Chicago Press.

Murray, F. and S. Stern. 2004. "Do Formal Intellectual Property Rights Hinder the Free Flow of Scientific Knowledge? An Empirical Test of the Anti-Commons Hypothesis." In *the 2004 International Industrial Organization Conference*. Chicago.

Organisation for Economic Co-Operation and Development (OECD). 2003. *Turning Science into Business: Patenting and Licensing at Public Research Organisations.* Paris, France: Organisation for Economic Co-Operation and Development.

Owen-Smith, J. and W. W. Powell. 2001. "To Patent or Not: Faculty Decisions and Institutional Success at Technology Transfer." *Journal of Technology Transfer* 26:99–114.

Rosenberg, N. and R. R. Nelson. 1994. "American Universities and Technical Advance in Industry." *Research Policy* 23:323–348.

Stokes, D. 1997. *Pasteur's Quadrant: Basic Science and Technological Innovation.* Washington, D.C.: The Brookings Institution.

The Chinese Statistics Bureau. 2001. "National R&D Expenditures in 2000." http://www.stats.gov.cn/tjgb/rdpcgb/qgrdpcgb/t20020331_15466.htm.

Thursby, J. G. and M. C. Thursby. 2002. "Who is Selling the Ivory Tower? Sources of Growth in University Licensing." *Management Science* 48:90–104.

U.S. Patent and Trademark Office (USPTO). 2002. "U.S. Colleges and Universities-Utility Patent Grants, Calendar Years 1969–2000. "

Walsh, J. P. and W. Hong. 2003. "Secrecy is Increasing in Step with Competition." *Nature* 422:801–802.

Young, T. A. 1999. "Personal Impressions of China." *The Journal of the Association of University Technology Managers* XI. http://www.autm.net/pubs/journal/99/china.html.

3 ICTs and the Human Body: A Social Representation Approach

Alberta Contarello and Leopoldina Fortunati

Rationale

There have long been discussions on Information and Communication Technologies (ICTs), their spread, and their reception in everyday life by different groups within the overall population. There have also been discussions on the artificialization of the human body (Katz 2003), and on the existence of "natural" and artificial technology (Fortunati 2003a, 2003b). It would appear that not only are New Technologies getting closer to the human body, but also that the latter is being penetrated by the former.

Among communicative technologies, we consider the mobile phone and the Internet emblematic of the ever-growing importance of mediated communication. The mobile phone, which unlike other diagnostic or therapeutic technologies, is at present getting closer and closer to the human body, has achieved trans-generational and mass diffusion. Moreover, it is linked to the expansion of communicative needs and expectations and "lives" in direct contact with the human body. The Internet, with its massive spread in the organization of knowledge, work, and health (Rheingold 1993; Castells 1996–1998; Rice and Katz 2001; Wellman 2001; Katz and Rice 2002), seems on the one hand to place the human presence in the foreground and, on the other, to modify perceptions of the body and of its presence (Wallace 1999).

The computer, the Internet, and the mobile phone have entered our everyday lives, our habits, speech, and conversations with remarkable rapidity. The combination of the mobile phone with the Internet will perhaps further strengthen this extension of communicative capacity. Yet, at the same time, it constitutes a greater artificialization of the human body and the much-discussed mediatization of the concept of self (Gergen 1991), mainly in its personal and relational aspects. In this convergence of technologies and the human body, an important encounter takes place, on the one hand, with the complexity of the fashion system (Fortunati *et al.* 2003) and, on the other, with the regimen of mobility, which is highly developed in our society (Ling and Haddon 2003).

There is now a considerable corpus of research aimed at understanding the relationship between technologies and concepts such as self, presence, and contact. The initial concern was with the role of the new ICTs in social life, especially with regard to the Internet (Döring 1999). Later, a growing number of scholars began to consider the particular modes of interaction between social actors and technological tools (Nyiri 2003; Kim 2004), either within a cultural psychology framework praising ethnomethodology (Mantovani 2002) or, within a social psychological perspective, moving from the search for simple main effects to more sophisticated and complex analyses (Bargh 2002) and to the relation between personality and the Internet (Hamburger 2005). Various theoretical models have been suggested to enhance the interactive and constructive nature of technology and social life (Joinson 2003; Flick 1995). Social psychological knowledge of these phenomena is expanding at the theoretical and empirical levels, as well as in terms of the use of ICTs in everyday life (Chiaro and Fortunati 2002).

In the present research we intend to investigate new information and communication technologies in relation to the human body. At a cognitive level, how do individuals deal with new technologies (Livingstone 2002), and in particular the mobile phone-Internet combination, as regards the human body (Fortunati and Contarello 2002; Contarello *et al.* 2003)? Is there concern and resistance to this much-discussed invasion of the body by the inorganic? Or is it welcomed into the sensibility of everyday life as a new opportunity? These are the questions we shall try to answer here.

Aim and Method

As previously mentioned, the aim of the present research is to monitor how the new technologies — mainly the mobile phone and the Internet — and the human body and the self are considered in this highly critical phase of what has been termed "mass prosthetization." To this end, the theoretical framework offered by Moscovici's (1961/76) social representation theory, appears suitable. Social representations are defined as forms "of knowledge which is socially elaborated and shared, and which has the practical aim of constructing a common, social reality" (Jodelet 1989:36). Not every item of knowledge is a good candidate for social representation. There has to be a *problem*, *contextualized* in a specific framework, and widely *shared* (Farr and Moscovici 1984). There has also to be *polymorphism*, with different aspects and voices given by different groups, so that the "object" has a specific value to specific groups (Moliner 1993). In our case, we explore how different approaches to the access and use of the Internet, mobile phones, and computers; as well as the different importance given to body-to-body presences and different social networks, interact with an emerging view of these technologies.

We thus try to reconstruct the social representations of ICTs,[1] in particular the mobile phone and the Internet, and of the human body. We also explore the social co-construction of the "technological" body that is emerging nowadays. For reasons of space, we will not deal with related topics — i.e., fashion and travel — which will be considered elsewhere. In previous studies, we examined relations between new communicative and therapeutic technologies and various features concerning the human body, communication, and emotions (the web, computers, mobile phones, machines, therapy, the human body, prostheses, communication, anxiety, power, intimacy, myself) and the possible convergence between the mobile phone and the Internet (Fortunati and Contarello 2005).

In the present work, we explore the representational field relative to New Technologies and to the Human Body. In particular, we focus on the different components of the representation — information, attitude, representational field — adopting a qualitative-quantitative methodology (Moscovici 1961/76; Le Bouedec 1984).

In order to study the first and the third component (information and representational field), we monitored the symbolic and emotional texture relative to the mobile phone, the Internet, and more generally the new technologies, on the one hand, and to the human body, on the other, via free associations (Di Giacomo 1980; Le Bouedec 1984). In order to measure attitudes towards the mobile phone, the Internet, and the computer, we collected answers to a Semantic Differential scale prepared for our purposes. We also gathered information through specific questions on practices relative to the mobile phone and the Internet and examined opinions on the relationship between these two technologies and the human body. The aim was to detect possible concerns, discomfort, or resistance that the respondents might express, or what kind of enthusiasm is triggered by this process of approaching the body.

[1]An increasing corpus of research is developing within this framework: Flick (1995); Capozza *et al.* (2003); Fortunati and Manganelli (in press), Contarello and Sarrica (2005); Contarello *et al.* (2004).

Participants and Procedure

Two hundred and eighty undergraduate male and female students from various faculties (Psychology, Engineering, Pharmacy) of Padua University in Italy took part in the study. They were enrolled in Psychology (43%), Engineering (28.6%), and Pharmacy (27.9%). Forty percent were men and 60% were women. The data were collected in the spring of 2002.

Our instrument consisted of five sections. First, we gave our participants a free association task, with six stimuli: the Internet, computer, mobile phone, journey, fashion, and myself. *(Write spontaneously the images and ideas which come to your mind...)*. Then, the participants were asked about two further stimuli, the new technologies and the human body, and responded to the following prompts: they are; they are the opposite of; they are similar to; they might be. These two stimuli were presented one per page in diagrammatic form. A third section contained a semantic differential to measure the participants' attitudes towards the mobile phone, computer, and the Internet. This seven-point scale was composed of four bipolar couples derived from Osgood *et al.*'s research (1957) and the Capozza's Italian adaptation (1977) that measure general evaluation (desirable-undesirable, agreeable-disagreeable, worthwhile-worthless, pleasant-unpleasant), and by 13 further pairs taken from a pre-test carried out with 20 students (who were asked to associate adjectives, nouns, or brief phrases to the three stimuli; the most relevant features were then retained and presented in a bipolar scale format). Fourth, a body investment scale (Orbach and Mikulincer 1998) was included to measure the importance given to Body image, Comfort with touch, Body care, and Body protection. This five-point scale, originally created to measure self-destructive behaviour, has also been used to measure normal conduct. Finally, the participants filled in the section of the questionnaire monitoring ownership, familiarity, and uses and practices with regard to the mobile phone, computer, and the Internet. Gender, age, education, and residence were the socio-demographic variables asked for, as well as information on the participants' social networks and status of relationships with others. These data were taken into account and entered into the analysis as illustrative variables in the representations in question.

Data Analysis

As mentioned above, we opted for a qualitative-quantitative approach, i.e., we first worked on the associated words, giving extensive space to the exploration of meanings conveyed by the respondents to the stimuli. Then, the distinct lexical forms associated with the stimuli were examined through analysis using the Spad.T package (Lebart *et al.* 1989). After reducing the synonyms and fixing the inclusion frequencies threshold, we scrutinized them following the Mocar procedure, to detect common words and words specifically associated with the mobile phone and the Internet. We then submitted the matrixes (participants × words, separately for each stimulus) to Correspondence Analysis (Benzécri *et al.* 1976), following the Aspar (participants × words for a single stimulus) and Aplum (stimuli × words) procedures. In the first case (Aspar), we explored the role of some illustrative variables — the use of the ICTs, the importance given to the human body and to body-to-body communication, the extension of the respondents' social network — upon the emerging representational fields. In the second (Aplum), we were able to analyze together the associations given to ICTs, the human body, and myself. The data from the semantic differentials were submitted to factor analysis, in order to discover the attitude and in particular the emotional temperature of the core stimuli (Internet, mobile phone, and computer).

Results

Ownership and Use of ICTs

Nearly all of the respondents (98.2%) owned a mobile phone and used text messages (97.9%). Most of them (85.4%) had a computer, and more than half claimed they used a computer daily (25.4%) or weekly (26.8%), but some (13.5%) declared they never used it. With regard to the Internet, its use was more varied (16.8% = daily, 29.6% = weekly, 26.8% = monthly, 18% = rarely, 9.3% = never), as was the use of email (5.4% = several times a day, 16.8% = daily, 35% = weekly, 18.2% = monthly, 20.4% = never, 4.3% = other).

Investment in the Human Body

The Body Investment Scale, submitted to exploratory factor analysis, showed the predicted pattern with only minor variations. On the whole, the participants appeared to be satisfied with their bodies (Mean = 3.83; median = 4.00), to like an average level of bodily contact during conversation (M = 3.22; m = 3.20), to take good care of their bodies (M = 4.09; m = 4.33), and to tend to avoid taking risks with it (M = 3.74; m = 3.67). These respondents, therefore, do not show the problematic and unsatisfied relation with their bodies that appears to emerge from the increasing resort, even by young people, to plastic surgery, diets, body modelling, and so on. With the split-half technique, we considered high and low scores on each factor (above and below the median), thus obtaining four new two-level variables.

As mentioned above, gender, ownership and the use of these devices, social networks and relationships, and body investment were entered as illustrative variables into the analysis (Aspar) regarding the Internet, the mobile phone, and the human body. We will first examine the most frequent lexical forms associated with the seven stimuli (which give us nuances of the "information" component as well as the emotional tone of the representation); second, the attitudes expressed towards the key stimuli; and third, the overall correspondence analysis involving the various stimuli (Aplum).

The "Content" of the Representation

The most frequent associations given to the proposed stimuli by at least 8% of the participants are reported in Table 1.

We found here that participants gave descriptions, mainly involving parts for the whole, when asked for free associations. The use of metonymy is widespread in all five stimuli, but with regard to the Internet, mobile phone, and computer, this kind of association takes up more space, leaving little room for evaluations and metaphors. The stimulus which prompted the highest number of evaluations was the human body, followed by journey. As regards Myself, a long list of lexical forms, mostly idiosyncratic, was produced; only in two of them — nice and shy — did we find some convergence among the participants. There were many other noteworthy aspects. One was the epithet

Table 1. Most Frequent Associations to the Proposed Stimuli

Internet	Internet-parts, computer, communication, information, omnipresence, science, usefulness, speed, fun, navigation, up-to-dateness, world, knowledge, notes, pornography, technology.
Mobile phone	Mobile-parts, communication, usefulness, friends, cost, handiness, telephone, annoyance, brands, reachability, fashion, fundamental, technology, contact, conversation.
Computer	Computer-parts, Internet-parts, fun, usefulness, work, technology, fundamental, science, speed, program, knowledge, difficulty, complexity.
The Human Body	Machine, exceptional, complex, perfect, beautiful, delicate, fundamental, unique, life, body-parts, physical, charming, identity, adaptable, functional, mysterious, looked-after, interest.
Myself	Nice, shy.
Fashion	Cost, clothes, shows, models, uselessness, conformity, color, elegance, appearance.
Journey	Fun, feast, holiday, sea, freedom, culture, sun, knowledge, friends, aeroplane, warmth, rest, discovery, relaxation.

"fundamental" attributed to the mobile and computer as well as, on different grounds, to the human body.

Attitudes

The items included in the attitude scale were submitted to exploratory factor analysis in order to single out the main thematic nuclei. Five factors emerged, accounting for 54.78% of the total variance. The first factor refers to *Appreciation and General Evaluation* (agreeable, amusing, pleasant, desirable, alpha = .81): all three technologies received positive evaluations as shown by the composite factor scores (Internet = 5.11; mobile = 5.03; computer = 5.09).

The further factors point to *Specific Evaluations,* contrasting information versus communication (the second), enhancing promptness (the third), convenience (the fourth), and artificiality (the fifth). However, as the internal coherence of these four factors did not prove to be satisfactory (alpha <.60), only the general appreciation measured through the first factor was entered in the overall analysis.

Representational Field

In order to explore convergences and divergences among the proposed stimuli, we performed a correspondence analysis (following the Aplum procedure), after a preliminary treatment of the data.[2] In this analysis (on a 5 × 184 table), the first factor opposes the mobile phone, Internet, and computer, on the one hand, to the human body and myself, on the other; the second opposes the Internet and the computer to the mobile; the third the human body to myself; and the fourth the Internet to the computer.[3] On the base of the associations

[2] It is an equivalence procedure, in which we put together different forms — singular, plural, and synonyms; and a threshold procedure, whereby we only retained lexical forms with at least five frequencies.

[3] A preliminary Aplum, including associations to journey and fashion, shows that the first factor opposes journey, and to a lesser extent the Internet, on one pole; to fashion, and the human body, on the other, evoking *Freedom* versus *Constraint.* The Internet shows a minor contribution to the factor (absolute contribution = 3.70), sharing with the stimulus journey associations such as discovery, fun, and knowledge; while the mobile phone does not enter into the definition of the factor. The second dimension opposes journey (and to a lesser extent, the human body) to the mobile and the Internet, with associations reminding us of a *Natural* versus *Technological* distinction. While realizing that interesting but minor connections pertain to the worlds of ICTs, mobility, and fashion in the view of our respondents, we focused specifically on the three core stimuli of the present research — the Internet, mobile phone, and computer — in relation to the human body and oneself.

produced (cf. Table 2), we propose to name and interpret the factors as follows:

1. *Communication Tools in Everyday Life* versus *The Machine of Marvels*. The factor opposes the three technologies under examination to the human

Table 2. Correspondence Analysis: The Field of Representation

First factor (I. = 36.97%)							
Communication tools in everyday life: factual considerations				**The machine of marvels: evaluative remarks**			
	coord.	**c.a.**	**c.r.**		**coord.**	**c.a.**	**c.r.**
Mobile phone	−.62	13.61	.22	Human Body	1.63	68.75	.89
Internet	−.43	7.26	.19	Myself	.86	6.66	.11
Computer	−.33	3.72	.11	Machine	1.73	8.07	.84
Mobile-parts	−.73	4.73	.20	Exceptional	1.88	6.91	.84
Computer-parts	−.45	2.01	.22	Perfect	1.95	4.73	.84
Internet-parts	−.48	2.58	.22	Beautiful	2.51	3.88	.97
Communication	−.48	1.58	.83	Delicate	1.95	3.78	84
Useful	−.43	1.16	.47	Complex	1.16	3.35	.94
Cost	−.71	1.08	.23	Unique	1.85	3.25	.95
				Human Body-parts	1.80	2.90	.82
				Life	1.57	2.83	.98
				Physicality	1.95	2.27	.84
				Charming	1.80	2.12	.98
				Well-groomed	4.51	1.89	.84
				Adaptable	1.73	1.65	.84
				Functional	1.73	1.65	.84
				Soul	1.84	1.53	.96
				Well-being	1.57	1.48	.96
				Expression	1.95	1.32	.84
				Identity	1.82	1.29	.53
				Interest	1.11	1.24	.83
				Dynamic	1.82	1.15	.98
				Mysterious	1.95	1.14	.84
				Device	1.23	1.05	.76

Second factor (CI. = 65.22%)								
The positive world of information					**The ambivalent support of relationships**			
	coord.	**c.a.**	**c.r.**			**coord.**	**c.a.**	**c.r.**
Internet	−.65	21.61	.43		Mobile	1.15	61.53	.78
Computer	−.59	15.30	.34	VS	Mobile-parts	1.47	25.13	.80
Internet-parts	−.83	10.14	.67		Cost	1.29	4.68	.77
Computer-parts	−.77	7.72	.76		Friend	1.00	3.44	.72
Information	−.80	2.48	.34					

Table 2. (*Continued*)

	coord.	c.a.	c.r.		coord.	c.a.	c.r.
Fun	−.58	1.92	.33	Convenience	1.05	3.03	.70
Science	−.71	1.47	.75	Brand	1.34	2.48	.78
Work	−.56	1.04	.21	Fashion	1.57	2.43	.82
Omnipresent	−.73	1.01	.32	Reachable	1.44	2.43	.79
				Telephone	.99	2.19	.63
				Annoyance	1.14	2.13	.77
				Conversation	1.57	1.78	.82
				Status-symbol	1.57	1.46	.82
				Noise	1.57	1.13	.82

Third factor (CI. = 87.98%)

The "Material" self				Personal and social identity			
	coord.	c.a.	c.r.		coord.	c.a.	c.r.
Human Body	−.52	11.40	.09	Myself	2.45	86.85	.89
Machine	−.74	2.41	.15	Nice	3.32	17.02	.94
Exceptional	−.78	1.90	.14	Shy	3.72	13.48	.93
Perfect	−.79	1.27	.14	Student	3.72	6.74	.93
Delicate	−.79	1.02	.14	Moody	3.72	6.64	.93
				Disposable	3.72	5.62	.93
				Sweet	3.72	5.62	.93
				Growing	3.72	5.62	.93
				Insecurity	3.08	4.62	.93
				Sensible	2.82	3.22	.84
				Curious	1.37	2.28	.71
				Family	2.23	2.02	.93
				Identity	.90	1.77	.44

Fourth factor (CI. = 100%)

Openness to the world				Technology for work and fun			
	coord.	c.a.	c.r.		coord.	c.a.	c.r.
Internet	−.61	43.64	.37	*Computer*	.73	56.01	.53
Communication	−.67	9.45	.63	Work	1.02	7.95	.67
Information	−.99	8.87	.51	Fun	.76	7.78	.56
Omnipresent	−1.01	4.59	.63	Software	1.38	5.29	.69
Up-to-dateness	−1.10	3.19	.53	Useful	.41	3.29	.44
Surf	−1.10	3.19	.53	Office	1.54	2.55	.73
Internet-parts	−.29	2.95	.08	Writing	1.54	2.55	.73
World	−1.02	2.57	.56	Help	1.23	2.31	.81
Pornography	−.84	1.40	.40	Computer-parts	.27	2.22	.08
Sex	−.99	1.35	.65	Fundamental	.47	1.95	.48
International	−1.11	1.32	.67	Technology	.50	1.72	.44
				Slowliness	1.14	1.39	.59
				Boredom	.86	1.25	.71

body and (to a lesser extent) the self, the communication tools mostly being described in their parts and components, and the human body referred to with a great variety of appreciative remarks. The particular blend of descriptions, evaluations, and astonishment associated with the human body evokes a powerful metaphor that seems to have its roots in the culture of classic Antiquity as well as of the Renaissance: it is the idea of a perfect machine. This radical opposition of the human body to the machines of communication and information is the point of departure for our entire analysis. Our respondents appear to advance the idea of irreconcilability: a divergence between the concept of communicative technology on the one hand and the human body and the self on the other. In other words, the body is perceived primarily as the opposite of technology, being experienced as the home of the self. The current debate on ICTs has revealed that many new technologies are converging on the human body for the purposes of prevention, diagnosis, and therapy; but also for communication, aesthetics, and fashion. This debate also registers a strong tendency to build a rhetorical stance that overestimates the fall of barriers and distinctions between the human body and technology. Within this debate, metaphors merge with arguments, but especially with empirical research data. By contrast, there emerges from this research project a basic distinction, in common sense terms, between ICTs and the human body. Whereas, to all intents and purposes, computers, mobile phones, and the Internet are seen as machines, the human body is only so on a metaphorical level. The metaphor only captures one of its numerous aspects. The human body is by no means contained within the definition of a machine; it merely includes it, as it includes many other images that serve to define it. For our respondents, the computer, mobile phone, and Internet are the product of science and technology; i.e., a human conquest that is described in its component parts and whose historical development can be traced. The human body refers instead to an idea of perfection, which deals with the transcendent: it is the idea of something marvellous that springs from classical antiquity, linked to the concept of wonder, a marvel of the gods, which causes wonderment.

It might be interesting to further specify how these stimuli — the ICTs and the human body/self — are conceptually integrated by our respondents. On the one hand, the ICTs are described in their parts and components rather than in their functions, affordances, and services, as if there were still a static, inert perception of the communicative instruments. This kind of perception is typical of children or novices who, in order to understand what a machine is and how it works (for example, a watch), have to take it to pieces and then re-assemble it. The practice of de-constructing technologies to better understand and define them probably implies that the object is still in the process of assimilation and domestication. On the other hand, the human body is described as a perfect, marvellous, exceptional, beautiful, charming, and unique machine. In particular, the body is perceived as well-groomed, adaptable, functional and dynamic; an instrument for the expression of one's own identity and interests. But it is also perceived as delicate, mysterious, and complex. This complexity is probably due to the fact that its physicality is regarded as the site of life, the soul and well-being, i.e., of immateriality. Thus, it is experienced as something that is not fully understood (mysterious) and so complex as to be delicate to manage. On the whole, the respondents do not appear to acknowledge the convergence of technologies on the human body that is taking place at present both in science and technology and in everyday life. They remain conceptually anchored to a sharp distinction between them.

2. *The World of Information* versus *The Ambivalent Support of Relationships.* The second factor distinguishes between ***communication***, linked to the mobile phone, with positive (friends, conversations, convenience) and negative features (cost, noise, annoyance); and ***information***, linked to the Internet and the computer, both encountered in the world of work and play. Our respondents perceive the information society as connected to science but also to fun in a world where computers are everywhere. It is interesting to note that, although the Internet is responsible for a great deal of interpersonal communication (for example, e-mail and chat), it is metabolized by our respondents under the umbrella of information. Instead, at the opposite pole, we find the mobile phone as the emblem of communication with friends (conversation: a.c. = 1.78, reachable: 2.43) and as a convenient device. At the same time, the mobile phone elicits the importance of brands and fashion. Through these two elements, it becomes a "status-symbol" in our respondents' perception. There is no lack of negative aspects, however: first of all the cost, but also annoyance and noise. Linked to the mobile phone is the cost of reachability, which is a problem for the individual's income. At the same time, the entry of the mobile phone into public space also implies negative features within the interpersonal framework. On the whole, information and communication are perceived as opposite spheres. However, the interconnection and hybridization between the worlds of communication and information have recently led to a specialist debate, while their conceptualization in common sense terms elicits two very distinct spheres. The mobile phone holds a particular position because of the respondents' references to costs, which are not only monetary but also social psychological, especially in terms of inconvenience and annoyance.

3. *The "Material" Self* versus *Personal and Social Identity*. In this case, the stimulus Myself gives the highest contribution to the factor and brings with it various associations that refer to personal (e.g., nice, shy, moody) and social (e.g., student, member of a group) identity, whereas the human body at the opposite pole is associated with an exceptional, perfect, and delicate machine. It is noteworthy that while the "material" self is seen as a perfect machine, with only positive connotations, the psychological self is regarded as a more fragile element, in which personal and social identity have positive aspects (nice, disposable, sweet, sensible, curious) mixed with negative ones, such as shyness, moodiness, and insecurity. Thus, we might conclude that human beings (or rather, our participants who are young and students) associate imperfection chiefly with the "immaterial."

4. *Openness to the World* versus *Technology for Work and Fun*. In this case, the opposition is between the Internet and the computer. The Internet seems to be linked not only to information but also to communication. Unlike the computer, communication is seen as specific to the Internet, evidently connected to e-mail and chatting, and it is regarded as having an international and world dimension. By contrast, the computer is considered useful and fundamental, linked both to work and fun, but is also seen as slow and boring. A great help in office life, the computer is particularly useful for writing (not yet for reading). Furthermore, it is seen as important more as software than hardware, as content rather than form, whereas the mobile phone is perceived as having a symbolic and aesthetic aspect. The computer is associated more with its function and performance than with symbolic and aesthetic values, which are attributed instead to the mobile phone — a key device for managing our relationships with others.

While the Aplum procedure enables us to consider the various stimuli together, the Asparm procedure allows us to explore in greater detail, individually, the core stimuli of the analysis — the Internet and the mobile — as well as our "contrast" stimulus — the human body.

The Human Body

To this stimulus 573 lexical forms were associated, 296 of which were distinct ones. After the equivalence and threshold procedures (retaining associations given by at least two respondents), 39 words were obtained, the most frequent of which were perfect (111 frequencies), machine (51), and unique (36). These words were entered in a 39 × 280 table. Figure 1 shows the first and second factors.

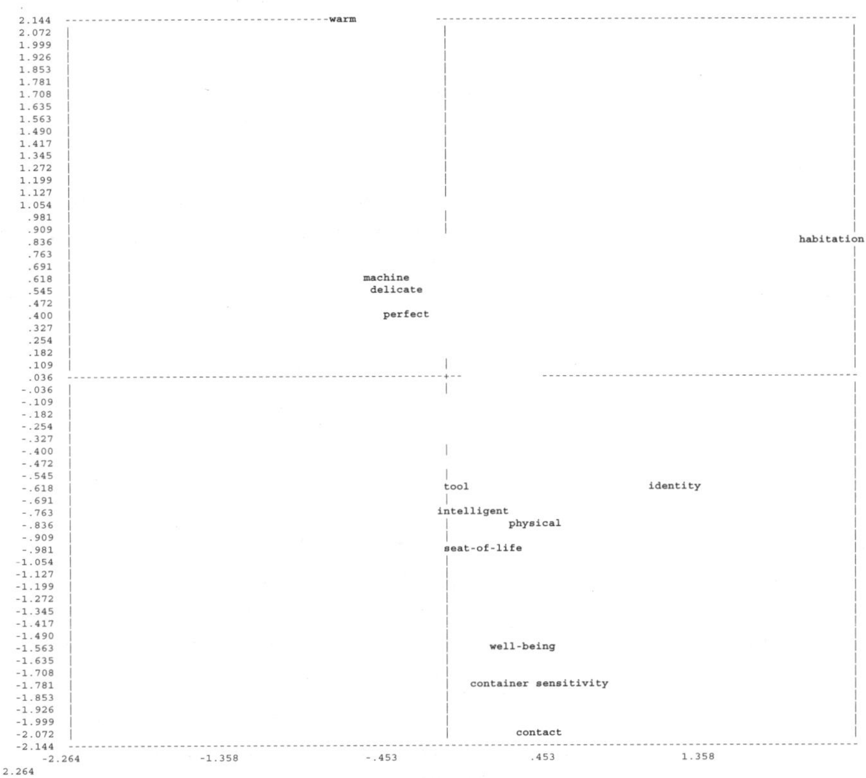

Figure 1. The Human Body. Correspondence Analysis: First and Second Factors

The first factor is well represented only at one pole, with the words habitation (absolute contribution = 79.4) and identity (5.8). These associations were mainly provided by students in Pharmacy, participants with a lower Body Image, and respondents who do not use a mobile phone or send text messages.[4] The second pole is "naked"; we find in it perfect (1.7) machine (1.0), but with minimal contributions. Based on these associations, the factor might be called: *The Home of the Self*, which is a wonderful definition for the human body. The second dimension presents at the positive pole the terms habitation (11.2), machine (4.2), perfect (3.3), and warm (3.0); and at the negative pole the terms contact (30.0), sensitivity (17.2), well-being (5.5), container (4.2), and seat of life (3.9). The idea of the human body as a *Perfect*

[4]Following Lebart and Salem (1988), only illustrative variables with Value-test > !2! are discussed.

Machine comes mostly from men, students who use the PC and the Internet daily, use email weekly, and have a higher Body image. By contrast, the idea of a *Means of Contact* is mainly given by women, students who rarely use a PC or the Internet, use email either very frequently (daily) or never, and declare a lower Body Image. The gender difference is not surprising, revealing the inclination of women towards contact and relations (cf. Tanner 1990; Wood 2001). An interesting result suggests a U-shaped relation between the use of the Internet as a communicative tool (via email) and views of the human body. Participants who stress the primacy of contact for the human body (the negative pole of the second factor) either use the Internet a great deal (daily) or never. By contrast, those who consider the body to be the perfect cosy "home" of the self report average use of the Internet as a communicative device.

The Mobile Phone

Participants associated 819 lexical forms with the stimulus mobile phone, 292 of which were distinct, the most frequent being text messages (112 frequencies), communication (48), useful (33), and friends (31). After the equivalence and threshold procedures mentioned above, the resulting matrix 40×280 was submitted to a correspondence analysis. The first three factors are considered here (Table 3).

The first factor contrasts *descriptive* notations (brand names, ring tones, text messages) with *evaluative* contradictory ones (useless, useful, awkward, stupid, convenient). The former are given mostly by participants who: report medium use of the mobile phone for text messages and conversation; consider it possible to communicate emotions via mobiles; and are students in Psychology. Evaluative comments are expressed by participants who rarely send text messages; never use the mobile phone for conversation; consider it impossible to communicate emotions via mobile phones; reveal lower standards of body care and protection; and are enrolled in Engineering. The second factor counterbalances the ambivalent characteristics of this technology, which we could call "the price to be paid," and the useless presence of brands in mobile telephony. The first pole is characterized by the annoyance that people have to suffer for their reachability; the second features a strong feeling of the uselessness of mobiles brands (but also telephone companies that provide services) and awkwardness, stupidity, harmfulness. From the illustrative variables (Table 3), there emerge various "patterns" based, first of all, on group membership: Engineering students are more likely to see both the positive and negative aspects of the technology, whereas Psychology students are more concerned with its weaknesses and dangers. It is interesting to note that future engineers, i.e. those who will design and construct the technologies, appear to be keener to see the machine in its ambivalent practical reality, while the future psychologists seem more critical towards machines, as well as towards the marketing techniques applied to them. Different practices of use are also important in

Table 3. The Mobile: Correspondence Analysis

First factor							
Description				**Evaluation**			
	coord.	**c.a.**	**c.r.**		**coord.**	**c.a.**	**c.r.**
Brands	−2.27	13.4	.19	Useless	4.02	12.9	.21
Telephone companies	−2.78	7.7	.12	Awkward	3.04	7.4	.11
Ring	−1.06	6.9	.13	Stupid	2.94	6.9	.12
Sms	−.47	5.1	.16	Useful	81	6.0	.13
				Fashion	1.06	4.5	.08
				Convenient	.76	3.9	.07
				Slavery	1.10	3.4	.07
Psychology				*Engineering*			
Use of mobile for conversation				*No use of mobile phone for conversation*			
Medium use of sms				*Low use of sms*			
Think it possible to communicate (+/−) emotions via mobile phone				*No (+/−) emotion in communication*			
				Low body care			
				Low body protection			
Second factor							
Pros: The price to be paid				**Cons: A useless impediment**			
	coord.	**c.a.**	**c.r.**		**coord.**	**c.a.**	**c.r.**
Annoying	−1.11	9.6	.15	Useless	5.36	24.7	.36
Reachable	−.74	2.5	.04	Brands	1.89	10.0	.13
				Awkward	3.40	9.9	.14
				Stupid	3.19	8.8	.14
				Telephone companies	2.33	5.8	.08
				Harmful	1.29	4.7	.07
Engineering				*Psychology*			
No mobile phone and sms				*Use of mobile phone for conversation*			
No use of mobile phone for conversation				*Communication of (+/−) emotions via mobile phone*			
No communication of (+/−) emotions via mobile phone				*Low comfort with touch*			
High comfort with touch							

Table 3. *Continued.*

Third factor							
Practicality				Status symbol			
	coord.	c.a.	c.r.		coord.	c.a.	c.r.
Practical	−4.73	21.2	.33	Status-symbol	1.96	8.2	.10
Relations	−3.95	18.5	.27	Awkward	2.68	6.8	.09
Fashion	−1.09	5.6	.08	Annoyance	.56	3.7	.05
Small	−1.65	5.2	.07	Useless	1.69	2.7	.04
Indispensable	−1.19	4.7	.07				
Speedy	−1.95	4.5	.07				
Battery	−1.66	2.6	.05				
Low use of the mobile phone				*Medium or no use of mobile phone*			
Don't often travel				*No positive emotion*			
No use of the mobile phone for work				*Low comfort with touch*			
				Pharmacy			
				Low body image			
				Possess no mobile phone			
				No use of sms			
				Often travel			
				Medium use of the mobile phone			
				Use of the mobile phone for work			

this connection. Participants who do not use the mobile for conversation, and those who do not rely on it to convey emotions mostly mention the reachability offered by the tool (although this is sometimes disturbing); while those who chat and believe that it is possible to communicate emotions through it stress the limits (awkwardness, stupidity, and uselessness) of the technology. An interesting result pertains to the Comfort with Touch subscale of the Body Investment scale: participants with higher scores on it stress the reachability that the means provides (probably seen as a tool to make contact and thus to help engage in body-to-body interactions), while those with lower scores complain of its uselessness without acknowledging its potential for "keeping in touch at a distance" with their partner. The third factor contrasts the *practicality* of the technology (practical, indispensable, speedy), particularly with regard to the world of interpersonal relations, to its *symbolic* (and annoying) aspects. In this case, we again find a U-shaped relation between the uses and views of the mobile phone: the practical and indispensable aspect of the technology is expressed by those who use it at a low level and by those who do not use it for work, while both those declaring a medium/high use and no use of a mobile phone refer to the status-symbol aspects.

The Internet

The Internet evoked 840 words, 280 of which were distinct from each other. After the usual treatment of the words, the most frequent categories being communication (71), computer (66), and information (44), the resulting 47 × 280 matrix was submitted for a correspondence analysis. The first four factors can be interpreted as follows (Table 4).

First, we find an opposition between a dynamic ***description*** and diversified ***evaluation***, the former due to participants with a higher use of the computer and the Internet and of their related functions, the latter with a lower use

Table 4. The Internet: Correspondence Analysis

First factor							
Description				**Evaluation**			
	coord.	c.a.	c.r.		coord.	c.a.	c.r.
Site	−1.41	11.8	.22	Useful	1.57	11.1	.18
Email	−.87	7.6	.18	Cold	2.31	7.7	.10
Computer	−.70	6.1	.15	Modern	1.58	5.9	.13
Computer-parts	−1.41	5.4	.09	Speedy	1.17	5.4	.11
Navigate	−1.25	4.5	.10	Infinite	1.33	3.9	.08
Chat	−.81	3.1	.07	Difficult	1.23	3.5	.07
				Curiosity	1.18	2.8	.06
				Dangerous	1.45	2.3	.05
Men				*Women*			
Use PC daily monthly				*Pharmacy*			
Use email				*Use PC rarely, never*			
Use email daily				*Use Internet rarely, never*			
Use chat				*No use of email*			
Use Internet daily				*Use monthly*			
Express (+/−) emotions via Internet				*No use of chat*			
				No utility			
Single				*No (+/−) emotions*			
Dates				*Engaged*			

Second factor							
Hot				**Cold**			
	coord.	c.a.	c.r.		coord.	c.a.	c.r.
Fun	−1.15	5.8	.08	Cold	5.98	62.0	.67
Porno	−1.58	4.9	.07	Work	1.68	4.3	.07
Chaos	−1.11	2.1	.03	Computer-parts	.94	2.9	.04
Men				*Women*			
Pharmacy				*Psychology*			
Use Internet weekly				*Use Internet rarely*			
Low body protection				*High body protection*			

Table 4. (*Continued*)

Third factor

Usefulness and danger	coord.	c.a.	c.r.	Productivity and destructiveness	coord.	c.a.	c.r.
Useful	−1.53	13.0	.17	Chaos	2.93	15.3	.20
Computer-parts	−1.41	6.7	.09	Cold	2.02	7.2	.08
Infinite	−1.39	5.2	.08	Porno	1.65	5.4	.08
Dangerous	−1.86	4.6	.07	Contact-world	1.47	3.9	.06
Modern	−1.15	3.8	.07	Work	1.53	3.6	.06
Colors	−1.75	2.7	.04	Global	.62	3.4	.06
Technology	−.94	2.2	.03	Information	.47	3.2	.07
				Freedom	1.53	2.6	.05
				Hackers	1.45	2.3	.05
Psychology				*Pharmacy*			
Use Internet monthly				*Use PC rarely*			
No use of email				*Use email*			
Express negative emotion				*No negative emotion*			
Often travel				*Don't often travel*			
Use email weekly				*Use Internet daily*			
Never use email				*Use email daily, monthly*			

Fourth factor

Sex and play	coord.	c.a.	c.r.	Study and contacts	coord.	c.a.	c.r.
Porno	−2.34	11.5	.15	Time	2.60	15.7	.23
Fun	−1.25	7.3	.10	Chaos	2.50	11.6	.14
Sex	−1.84	4.7	.07	Study	1.87	8.1	.11
Site	−.60	2.8	.04	Computer-parts	1.46	7.5	.09
Sms	−1.54	2.7	.04	Contacts	2.14	4.3	.06
Curiosity	−.97	2.4	.04	Freedom	1.60	3.0	.05
Cold	−1.05	2.1	.02	Far-away-places	1.68	2.6	.05
				Contact-world	1.14	2.4	.04
Men				*Women*			
Pharmacy				*Engineering*			
Use PC for writing and data processing				*Use PC weekly*			
Use PC to play				*No use for writing and data processing*			
Use Internet weekly				*No use to play*			
Use Internet for contacts				*Use Internet monthly*			
Use chat				*No chat*			
Use email daily				*Use email several time a day*			
Don't often travel				*High body protection*			
				Often travel			
				Engaged			

of the PC and the Internet. Second, there is a contrast between the Internet regarded as *cold* and a mixture of associations such as fun, porno, chaos that make us think of a *hot* side of the technology. The former is linked with rare use of the Internet and higher scores in Body Protection, the latter with more frequent (weekly) use of the technology and lower scores on Body Protection. Third, two different blends appear that recall some variations on a theme: on the one hand psychologists, participants who rarely (monthly) or never use the Internet and email, and who tend to share negative emotions via the Internet declare its *usefulness & danger*; on the other, pharmacists, participants using email, and those who do not share negative emotions via Internet mention its double-faced *productivity & destructiveness*. Fourth, a light *versus* committed dimension is suggested by associations that refer to *sex and play* on one the hand and *study and open (or free) contacts* on the other. The former is characterized by a medium to high use of the PC and Internet, in its various functions; the latter by a very low (monthly) or quite high (several times a day) level of access to the Internet. This second pole is the expression in particular of students in Engineering, participants who are engaged, and those with higher scores on the Body Protection scale. Moreover, in three of the four factors, gender proves to be an important variable: men are found on the descriptive pole of the first factor, the hot pole of the second, and the sex and play of the fourth; whereas women more often resort to evaluation, mention coldness and refer to study and contacts with the world. Thus, women's perception and views of the Internet would appear to be opposite to men's. Although several research studies in the last few years have shown an increasing interest among women in pornography (Asper 1999) and videogames, this is not supported by the present research.

Discussion

From the collected data a rich and varied picture emerges. As regards the *information* component of the representation, we find that:

a. The Internet, mobile, computer, as well as the other items selected are mostly considered useful, described in their parts, and are less often mentioned in metaphorical terms. On a closer reading, however, several interesting aspects arise, the most striking of which is the ambivalent attitude to these technologies.
b. The Human Body appears to be seen as perfect, wonderful, unique, a splendid "machine," and as being delicate but complex.

c. The self elicits a rich variety of associations, some of them pertaining to social and relational aspects, but most of them referring to idiosyncratic psychological features. It is the element that shows particular fragility with respect to the material part — the human body — which contains it.

As regards the second component, we see that *attitudes* towards the Internet, the mobile phone, and the computer show an overall positive evaluation, with differences in more specific aspects.

The structural analyses aimed at mapping the *representational field* show a sharp distinction between the *description and evaluation* of the proposed stimuli. On closer scrutiny, various organizing principles emerge regarding the various stimuli:

Along the poles of the various dimensions, different groups (as regards location, practice, and concerns with the human body) take different positions. Interestingly, some of them trace a highly coherent framework. As regards gender, women see the human body as a means of contact; they perceive the Internet as useful, but also cold, and mention its use for study and contact with the world. Men, on the contrary, tend to describe various aspects of the Internet, mentioning its "hot" features linked to sex and play. No important difference appears in attitudes toward the mobile phone. As for the faculty affiliations of the respondents, students of Pharmacy show a particular view of the human body and stress the exciting but also potentially destructive facets of the Internet; students in Engineering give evaluations (mostly negative) of the mobile phone, acknowledging the reachability it provides for its users, and stress the expansion of time and space as a result of the web (linking the Internet to study and contact with the world); students in Psychology mostly describe the mobile by emphasizing its negative features, as well as showing some concern about the twofold nature of the Internet as both useful and dangerous.

Another interesting point is emotional investment. Investment in the human body shows some interaction with the participants' view of the body itself (considered a perfect machine by individuals with a high body image, but a means of contact by those with a low body image). Regarding the mobile phone, its role of reachability is acknowledged by those with a high degree of comfort with touch in interaction, while those less at ease with touch consider the mobile a useless impediment. As for the Internet, participants with higher body protection stress the cold side of the technology, and its usefulness for study and contact with the world; while those with a lower score on that factor mention the "hot" side of the web.

A scrutiny of the other illustrative variables examined indicates that different practices play a considerable role in the representation of ICTs. A medium level of use is often linked to a richer and more detailed description of ICTs, while both rare and very frequent use lead to (ambivalent) evaluations. The different ways in which the technologies are used (for work, play, making contact; through text messages, email, data processing, information gathering...) also play a role, indicating different positions and views on the topics investigated in relation to the respondents' various practices.

Conclusion

The theoretical framework and methods we proposed appear to be appropriate for exploring not only the uses, but also the images of and attitudes towards ICTs. Our findings make a contribution to the broader study of the social representations of ICTs, providing interesting considerations on how they are evaluated, judged, and described. More recent data (Contarello *et al.* 2004), however, show that the picture is changing fast, thus increasing the need to monitor this active phase in the development of ICTs and their perceptions (Flick 1995).

In light of our findings, the answers to our initial research questions can be summarized as follows:

a. Though rich and varied, "social thought" does not seem to have integrated at a conceptual level the ongoing process of convergence between the human body and the technologies. This is perhaps because of the emergence of a clear opposition between the human body and the technology. There are neither straightforward rejections, nor great enthusiasm as regards this invasion of the body by the inorganic. The very existence and development of this process are going unnoticed by our respondents, but this is perhaps due to their youth.
b. The net distinction between communication and information, perceived as distinct and opposite spheres, is another striking result, since information and communication are certainly merged in the Internet and in the mobile phone. The fact that this hybridization is not perceived may be because the process is very recent. Alternatively, it might mean that young people are unaware of the convergence between these communicative spheres. Further data are needed in order to clarify this issue.
c. Lastly, this research project confirms once again the positive attitude towards technologies on the affective plane, although there is some concern

as to specific devices. The Internet is perceived as having a twofold nature: work and play, hot and cold, science and chaos. The mobile phone is associated with a rising "price-to-be-paid," both because of the pointlessness of mobile-phone brands and the way it disturbs social space.

Further research is required using broader samples from different age groups, and educational and social contexts, as it is still unclear whether the youth of our respondents is affecting their answers or whether the "social thought" we have attempted to give voice to is still at an early stage of elaboration.

References

Asper. 1999. *Pianeta giovani. Roma:* Associazione per lo Studio dell'Analisi Psichica e la Ricerca in Sessuologia.

Bargh, J. A. 2002. "Beyond Simple Truths: The Human-Internet Interaction." *Journal of Social Issues* 58:1–8.

Benzécri, J.-P. *et al.* 1976. *L'analyse de Données. Vol.2: L'analyse des Correspondances.* Paris: Dunod.

Capozza, D. 1977. *II Differenziale Semantico.* Bologna: Pàtron.

Capozza, D., R. Falvo, E. Robusto and A. Orlando. 2003. "Beliefs about Internet: Methods of Elicitation and Measurement." *Papers on Social Representations* 12: 1–14.

Castells, M. 1996–98. *The Information Age: Economy, Society, and Culture*, 3 vols. Oxford: Blackwell.

Chiaro, M. and L. Fortunati. 1999. "Nouvelles Technologies et Compétences des Usagers." *Réseaux* 17(96):147–182.

Contarello, A. and M. Sarrica. (2005). "ICTs, Social Thinking and Subjective Well-Being. The Internet and Its Representations in Everyday Life." *Computers & Human Behavior.* Available online at: http://www.elsevier.com/wps/find/journaldescription.cws_home/759/description#description.

Contarello, A., L. Fortunati, P. G. Fernandez, E. Mante, O. Versinskaya and D. Volovici. 2003. "Social Representations of ICTs and the Human Body. A Comparative Study in Five Countries." In *The Good, the Bad and the Irrelevant: The User and the Future of Information and Communication Technologies*, edited by L. Haddon, E. Mante-Meijer, B. Sapio, K.-H. Kommonen, L. Fortunati and A. Kant. Helsinki: Media Lab UIAH.

Contarello, A., L. Fortunati and M. Sarrica. 2004. "*Pensiero Sociale e Nuove Tecnologie.* II Contributo delle RS allo Studio del Cambiamento Sociale Indotto Dagli ICTs." Italian Association of Sociology (AIS), Methodology Section: *Lo Studio del Mutamento Sociale: Problemi Metodologici e Ricerche Empiriche* 28–29 Gennaio, Salerno.

Di Giacomo, J.-P. 1980. "Intergroup Alliances and Rejections within a Protest Movement. Analysis of the Social Representations." *European Journal of Social Psychology* 10:329–344.

Döring, N. 1999. *Sozialpsychologie des Internets.* Göttingen: Hogrefe.

Farr, R. and S. Moscovici, eds. 1984. *Social Representations.* Cambridge: Cambridge University Press.

Flick, U. 1995. "Social Representations." In *Rethinking Psychology*, edited by J. A. Smith, R. Harré and L. van Langenhove. London: Sage.

Fortunati, L. 2003a. "The Human Body: Natural and Artificial Technology." In *Machines that Become Us*, edited by J. E. Katz. New Brunswick, New Jersey: Transaction Publishers.

Fortunati, L. 2003b. "Real People, Artificial Bodies." In *Mediating the Human Body: Technology, Communication and Fashion*, edited by L. Fortunati, J. E. Katz and R. Riccini. Mahwah, New Jersey: Erlbaum.

Fortunati, L. and A. Contarello. 2002. "Internet-mobile Convergence: Via Similarity or Complementarity?" *Trends in Communication* 9: 81–98.

Fortunati, L. and A. Contarello. 2005. "Social Representation of the Mobile: An Italian Study." In *When Mobile Came: The Cultural and Social Impact of Mobile Communication*, edited by Shing Dong Kim. Seoul: Communication Books.

Fortunati, L., J. E. Katz and R. Riccini., eds. 2003. *Mediating the Human Body: Technology, Communication and Fashion.* Mahwah, New Jersey: Erlbaum.

Fortunati, L. and A. M. Manganelli. (In press). "The Social Representation of Telecommunications." *Personal and Ubiquitous Computing.*

Gergen, K. J. 1991. *The Saturated Self.* New York: Basic Book.

Hamburger, A. Y., ed. 2005. *The Social Net: The Social Psychology of the Internet.* New York: Oxford University Press.

Katz, J. E. and M. Aakhus., eds. 2003. *Perpetual Contact: Mobile Communication, Private Talk, Public Performance.* Cambridge: Cambridge University Press.

Katz, J. E., ed. 2003. *Machines that Become Us.* New Brunswick, New Jersey: Transaction.

Katz, J. E. and Ronald E. Rice. 2002. *Social Consequences of Internet Use. Access, Involvement, and Interaction.* Cambridge, MA: The MIT Press.

Kim, S. D., ed. 2004. *Mobile Communication and Social Change.* Seoul: Hallym University.

Jodelet, D., ed. 1989. *Les Représentations Sociales.* Paris: PUF.

Joinson, A. N. 2003. *Understanding the Psychology of Internet Behaviour. Virtual Worlds, Real Lives.* New York: Palgrave Macmillan.

Lebart, L., A. Morineau, M. Becue and L. Haeusler. 1989. *SystèmePortable pour l'analyse des Données Textuelles (SPAD.T).* Paris: Cisia.

Lebart, L. and A. Salem. 1988. *Analyse statistique des données textuelles.* Paris: Dunod.

Le Bouedec, G. 1984. "Contribution à la Méthodologie d'étude des Representations Sociales." *Cahiers de Psychologie Cognitive* 4:245–272.

Ling, R. and L. Haddon. 2003. "Mobile Telephony, Mobility, and the Coordination of Every Day Life." In *Machines that Become Us*, edited by J. E. Katz. New Brunswick, New Jersey: Transaction Publishers.

Livingstone, S. 2002. *Young People and New Media*. London: Sage.

Mantovani, G. 2002. "Internet Haze: Why New Artifacts Can Enhance Situation Ambiguity." *Culture and Psychology* 8(3):307–326.

Moliner, P. 1993. "Cinq Questions à Propos des Représentation Sociales." *Cahiers Internationaux de Psychologie Sociale* 20:5–14.

Moscovici, S. 1961/76. *La Psychanalyse. Son Image et Son Public*. Paris: PUF.

Nyíri, K., ed. 2003. *Mobile Democracy. Essays on Society, Self and Politics*. Vienna: Passagen Verlag.

Orbach, I. and M. Mikulincer. 1998. "The Body Investment Scale: Construction and Validation of a Body Experience Scale." *Psychological Assessment* 10(4):415–425.

Osgood, C. E., G. J. Suci and P. H. Tannenbaum. 1957. *The Measurement of Meaning*. Urbana: University of Illinois Press.

Rheingold, H. 1994. *The Virtual Community. Homesteading on the Electronic Frontier*. New York: Harper Perennial.

Rice, R. E. and J. E. Katz. 2001. *The Internet and Health Communication*. London: Sage.

Tanner, D. 1990. *You Just Don't Understand: Women and Men in Conversation*. New York: Ballantine Books.

Wallace, P. 1999. *The Psychology of the Internet*. Cambridge: Cambridge University Press.

Wellman, B. 2001. "Physical Place and Cyberplace: The Rise of Personalized Networking." *International Journal of Urban and Regional Research* 25(2):227–252.

Wood, J. T. 2001. *Gendered Lives: Communication, Gender, and Culture*, 4th ed. Belmont, CA: Wadsworth Publishing.

4 Technological Development and Society: The Discourse on PGD in Germany

Kerstin Wüstner

Introduction

Technology can only develop within a social system.[1] If one imagined a "global world," technological development would be quite similar in all countries. The example of Preimplantation Genetic Diagnostics (PGD) shows that the German government chooses not to allow this technology, while it is legal in most other European countries. This paper aims to analyze why this is the case, and to show how the discourse on PGD in Germany can be described. A sociological device to analyze discourses is the "discourse analysis." It focuses on how social construction processes can be reconstructed and how communication and legitimation processes of the structure and pattern of sense can be explained (Keller *et al.* 2001). Society constructs actuality (Berger and Luckmann 1967); this process is sustained by acting and interpreting. The way in which understanding is produced is subject to the group in which the discourse takes place. But before dedicating the discussion to discourse analysis, we will examine some aspects of technology and society.

Technology: Ultrasound, PD, IVF, and PGD

While the embryo was out of the range of direct inspection until the early 1960s, the first attempts to examine it in utero were undertaken by using

[1] In this paper, we will not differentiate between "technology" and "techniques," but we will refer to the term "technology." Technology is perceived to be an artificial construct, comprising a pool of practices, procedures, and instruments. Technological developments are supposed to be advantageous for society, although most could have non-intended or negative side-effects as well.

ultrasound from the mid-1960s onward. The use of early ultrasound only allowed for an analysis of the embryo's external appearance, but did not tell anything about its genes. In 1968 this changed, as amniocentesis was used for the first time in order to analyze the embryo's genes — thereby making Prenatal Diagnosis (PD) available. Before that, amniocentesis was a means for abortion by inserting saline solution in the utero (Holzgreve *et al.* 1995). Nowadays, physicians take a sample of amniotic liquid and analyze some of the embryo's cells, which can be found in the fluid. In the very beginning of genetic analysis, it was only possible to make a so-called karyogram in which the number of chromosomes[2] is displayed. Thus, this technique enables scientists to tell parents whether their baby has Down syndrome, for example, because of the surplus chromosome number 21. Also, the structure of the chromosomes could be analyzed to check whether or not the baby might be genetically "normal." Since the late 1980s, scientists have been able to analyze not only the number and structure of chromosomes (cytogenetic analysis), but also single genes (molecular analysis), which can be responsible for conditions such as cystic fibrosis.[3] Amniocentesis is usually applied in the 14th to 18th week of gestation. Although a new kind of quick analysis, fish-diagnosis,[4] now exists it still usually takes between two and three weeks until the genetic analysis is completed. Thus, a woman is in the 16th to 21st week of gestation when she receives the results. In the case of a pathological result, the parents have to come to a decision quickly, because the pregnancy is already in an advanced stage. A late abortion is a severe physical and psychological burden. If an abortion takes place in the 23rd week of gestation, the embryo might be able to live outside the woman's body, so it would be viable after the abortion.

Therefore, another technique was developed, the chorionic villi biopsy,[5] which can already be applied from the 11th week of gestation. Another advantage is that the genetic analysis does not require as much time. Yet the risk of induced loss of the baby due to the invasive technology of chorionic villi biopsy is higher (between 3% to 8%, e.g., Murken and Greve 1994; Cara 1992) than

[2]Chromosomes are DNA molecules. They are associated proteins that carry portions of hereditary information. Usually, every human has 23 paired chromosomes.

[3]Cystic fibrosis is a genetic disease. A defective gene causes the body to produce an abnormally thick, sticky mucus that clogs the lungs and leads to life-threatening lung infections. These thick secretions also obstruct the pancreas, preventing digestive enzymes from reaching the intestines to help down and absorb food (http://www.cff.org).

[4]Fish-diagnosis (= fluorescent in situ hybridization) is a new detection technique.

[5]Chorionic villi sampling is a method to withdraw a small amount of chorionic tissue (tissue from the developing placenta).

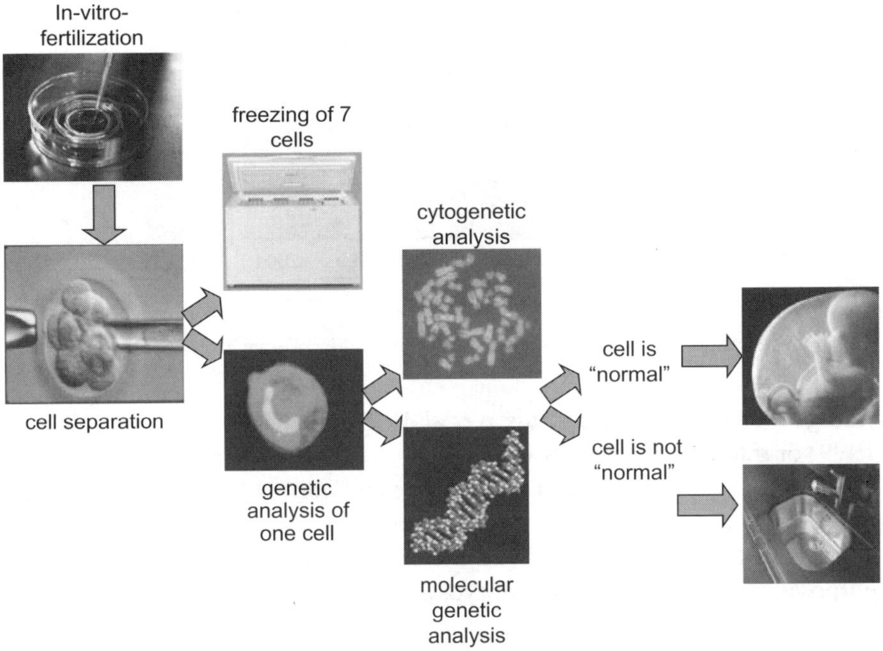

Figure 1. Simplified Scheme of PGD-steps

it is for amniocentesis (about 0.7% to 3.1%, e.g., Murken and Greve 1994; Holzgreve *et al.* 1995).[6]

In the 1970s, the artificial in-vitro fertilization (IVF) was developed, and the prenatal genetic diagnosis of artificially fertilized cells, PGD, was put into practice from the early nineties onwards (Handyside *et al.* 1990). Figure 1 illustrates steps of the PGD, referring to the example of an eight-cell stage analysis (sometimes cells at other stages are also used, beginning from the two-cell stage to the twelve-cell stage, e.g., Levinson *et al.* 1995, Munné *et al.* 1994) and an analysis of just one cell (sometimes two cells are extracted for genetic diagnosis).

The starting point of this process is an IVF. When all cells are still omnipotent, which means that they have not specialized yet, but are capable of potentially creating a human being, one of the cells is extracted and analyzed for

[6]Other prenatal testing methods such as the triple-test or fetoscopy also exist, but they will not be described here as the other techniques are more important in practice.

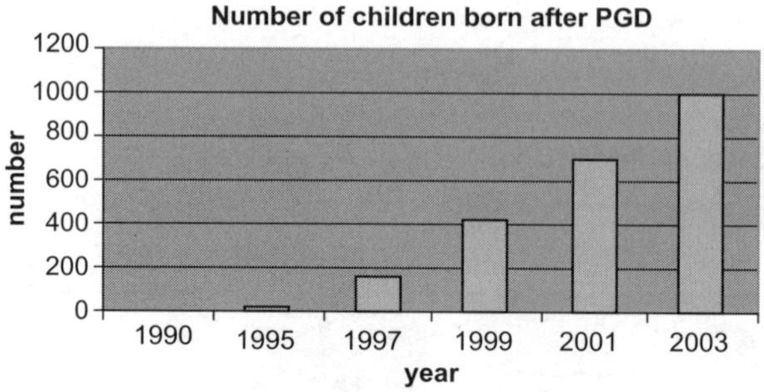

Figure 2. Trends in the Number of Children Born after PGD

sexing (Griffin *et al.* 1992; Staessen *et al.* 1999), aneuploidy,[7] (Gianaroli *et al.* 1999) or translocations[8] (Munné *et al.* 2000). The remaining cells are frozen at −196°C in fluid nitrogen. If the tested cell is "normal," some of the remaining cells will again be implanted in the woman. If the cell is not "normal" or if it reveals the "wrong" sex, all of the cells are destroyed or used for other purposes, such as research on embryos.

The number of centres that perform PGD has increased over the last few years, as have the number of children born following PGD. After the first PGD baby was born in 1990, five children were born in 1992, 31 by 1994 (Verlinsky *et al.* 1994), 34 by 1995 (Harper 1996), about 100 by 1996 (Verlinsky *et al.* 1997, Handyside *et al.* 1998), 424 by 1999,[9] 693 by 2001,[10] and about 1,000 by 2003.[11] Figure 2 illustrates this trend.

Statistical data about PGD provide very interesting insights. The European Society of Human Reproduction and Embryology (ESHRE) collected data from 25 European centres. Their last report (2002) refers to the period from 1994 to 2001 and includes 1,561 couples who wished to have PGD. Most of them, 96%, were referred to PGD due to genetic risks (chromosomal, X-linked, e.g., fragile X syndrome, and autosomal genetic abnormalities, e.g., Chorea Huntington disease), yet 30 clients (2%) wanted only social sexing. The women underwent 2,074 cycles in which nearly 27,000 eggs were retrieved. From the total number of oocytes collected, about 62% could be

[7]Aneuploidy means that the number of chromosomes is not normal. It is possible that there is an additional chromosome or a missing one.
[8]In the case of a translocation, pieces of some chromosomes have, for example, switched places.
[9]See http://www.spdfraktion.de/cnt/rs/rs_dok/0,20574,00.htm.
[10]See Bundestag Enquete-Kommission Medizin Drucksache 14/9020 S.84f.
[11]See http://www.mta-verband.at/zytogenetikforum/theorie/pid.htm.

fertilized, 48% were suitable for biopsy, 17% were suitable for transfer. A total of 309 pregnancies were achieved, with 215 births and a total of 279 babies (ESHRE 2002).

Almost half of all women wished to have a PD as well, in order to ensure that the PGD diagnosis was correct or because they wanted to use another instrument to ensure that the baby was healthy. In the described sample, seven false PGD diagnoses were revealed (about 3%), due to which four pregnancies were aborted. Also, there were nine cases of multiple pregnancies where 15 selective fetozides were carried out. All in all, a clinical abortion in the first trimester occurred in 11% of the cases. The study also provides data on the health of babies born following a PGD procedure: 6.6% had malformations, and 42% had neonatal complications, of which 4% ended in death (ESHRE 2002).

These data show that PGD is accompanied by some burden for the woman, as such the need for a high number of eggs, with PGD succeeding only in some cases. The data also underscore that PD is often used as an additional means of diagnosis. The number of complications in children born following the use of PGD is noticeably high.

Society and Technology: The German Situation

In Germany, it has become "normal" for (pregnant) women resort to different forms of technology to conceive and to check the health of the unborn.

If women have difficulties conceiving, they can refer to IVF. In 2001, 75,086 cycles were reported in Germany. More than 48,000 transfers of 110,000 embryos were made; 13,666 pregnancies resulted, and 7,062 children were born (Deutsches IVF-Register 2001).

The compulsory health insurance — through which approximately 88.5% of the German population are insured — bears the costs of (up to three attempts of) IVF if it has been established that this is the only way for a woman to conceive. In the case of private health insurance, whether or not the costs are covered depends on the specific contract. The costs for one IVF differ quite a bit, ranging from 3,500 to 7,000 Euros.

During pregnancies, technology has also achieved quite a lot of importance. For example, according to the German "Mutterschaftsrichtlinien,"[12] three ultrasound checks are recommended: between weeks 9–12, 19–22, and

[12]The "Mutterschaftsrichtlinien" are guidelines that list recommended medical procedures during pregnancy.

Table 1. PD in Germany, 1996–1999

	1996	1997	1998	1999
Live births	796,013	812,173	785,034	770,744
Amniocentesis & chorion biopsy	62,331	67,225	66,958	67,320
Loss due to invasive diagnostics	600	700	700	700

Source: Statistisches Bundesamt

29–32 of gestation. Women usually decide to take advantage of them. The cost of these checks is also covered by all health insurance companies, regardless of whether they are compulsory or private. If anything suspicious is found (e.g., the risk of abruption of placenta, foetal anomalies, or if there exists a family history that reveals genetic risks), more ultrasonic testing is offered and covered by insurance companies.

Besides ultrasonic testing, PD is legal and widely accepted in Germany. There has, however, been criticism of the procedure, mainly from specially constituted organizations (e.g., the "network against selection by PD"). But when couples are in a position to decide for themselves whether or not to resort to PD, most of them choose to use this diagnosis. About 80% of women to whom PD is offered decide in favour of it. When receiving a suspicious ultrasonic testing result, all women choose to have PD (Nippert 2001).

Several techniques are used: chorion biopsy, amniocentesis, biopsy of the umbilical cord, or foetal biopsy. From 1996 until 1999 the utilization of PD was as shown in Table 1.

Nearly 70,000 PDs are carried out each year. The main reason for women to choose this procedure is their age — women older than 34 are defined as "risky" cases and thus have to be informed about PD. If the physician neglects to inform the clients, they can sue for damages. There have been several court decisions in which physicians were order to make payments because they missed to inform their clients (sufficiently) about PD.

Abortions are legal in Germany under certain circumstances (§218 StGB).[13] First, a woman can abort the child during the first twelve weeks of gestation without needing to justify her motives for doing so. She only needs to attend a counseling session before the abortion. Second, an abortion is legal if there is evidence of a medical indication, which means that the woman would face unbearable burdens if she gives birth to the child. Thus, this indication can include social as well as health aspects. As there are no time

[13]§218 is an article in the German civil code governing abortion as described as above.

Table 2. Abortions in Germany, 1993–2000

	1993	1996	1998	2000
Live births	798,447	796,015	785,034	766,969
Number of abortions	111,236	130,899	131,795	134,609
Ratio of medical indications (in %)	6.26	3.68	3.29	2.70
Number of late abortions, after the 23rd week of gestation	90	159	175	154

Source: Statistisches Bundesamt 2001

limits anymore, it would be legal to abort a child until up to one minute before birth. Third, a criminological indication may be used to justify an abortion, e.g., when a woman wishes to abort a child she is expecting after a rape.

As the law in Germany no longer regulates time limits, physicians and midwives agree until to what point in time they will support abortions. Yet this only constitutes a social agreement. In the majority of cases, experts decide to support abortion only before the 23rd week of gestation, because from that time on, the baby is potentially able to live outside of the mother's body. If the aborted baby were to live, physicians would have the duty to care for it. That is why the foetus is sometimes killed in utero. This procedure is legal, whereas it would be considered murder if the baby was killed after the abortion.

Table 2 reflects the official numbers of abortions in Germany. Yet one should note that there exist a number of unknown cases, partly because some women go to other European countries, such as the Netherlands, where abortion regulations are assumed to be less strict and more informal.

It is apparent that the number of abortions has steadily increased over the last few years, whereas the ratio of medical indications has decreased. This means that most women decide to have an abortion due to social reasons in early pregnancy. Further, there are still cases of late abortions but the number is comparatively small. The German Society of Gynaecology and Obstetrics (2003) has questioned these numbers, however, as some late abortions are labeled "stillbirths."

Although different technological devices are used to assist pregnancy, one technology is excluded: PGD. This technology is strictly forbidden in Germany by law, in form of the so-called "Embryonenschutzgesetz"[14] which was passed 1991. This law states that embryos may not be cloned. As one

[14] "Embryonenschutzgesetz" is a law for the protection of embryos.

omnipotent cell can develop a human being, it must not be used for a genetic analysis. Further, embryos can only be produced, if and only if, the aim is a pregnancy. Yet another reason why PGD is illegal in Germany is that there exists the risk that the embryo will be destroyed if genetic mutations can be found in one of the cells. Interestingly, PGD has remained illegal in Germany although it is applied in most other European countries (see Table 3).

As PGD is not practised in Germany, "medical tourism" to undergo this procedure abroad has developed, especially to Belgium and the Netherlands (Krones *et al.* 2002). Several years ago, an ongoing, broad discussion about PGD emerged, with a number of voices claiming that the "Embryonenschutzgesetz" should be changed and PGD legalized.

Table 3. PGD Regulations within Europe and Some Countries of the Former Soviet Union

Countries in which PGD is legal or practised	Countries in which PGD could be practised but utilization is not possible for economic reasons	Countries in which the legal status of PGD is unclear and PGD is not practised	Countries in which PGD is illegal and thus not practised
Belgium	Latvia	Austria	Germany
Bulgaria			Switzerland
Czech Republic			
Denmark			
Finland			
France			
Greece			
Hungary			
Iceland			
Italy			
Lithuania			
Norway			
Portugal			
Russia			
Slovakia			
Slovenia			
Spain			
Sweden			
The Netherlands			
Ukraine			
United Kingdom			

No information could be gathered about countries not included in this list. In some rare cases information about PGD was not entirely clear.

Publications as a Mirror of Social Representations

In a next step, we will analyze how the discussion, or the discourse, about PGD can be described. In other words, the question is, how is PGD socially represented (Moscovici 2001)? In order to investigate this question, different documents will be analyzed. This enables us to include social representations of different social systems at the same time. An analysis of publications can highlight how certain topics are socially represented. Above, the way in which they are presented can have an effect on the perception of social groups and individuals in turn. Thus, publications are a result of a social construction process and they influence this construction process again.

The material used for the analysis consisted of newspaper articles published in the "Süddeutsche Zeitung," a daily newspaper for the general public. It is distributed all over Germany and has a daily print run of about 440,000. The "Süddeutsche Zeitung" contains a scientific column. Further, articles published in "DIE ZEIT" are analyzed, a weekly newspaper focusing on academics as their most important target group. Its print run is about 460,000. DIE ZEIT is also distributed without any regional restrictions. The third newspaper we will refer to is the "Berliner Zeitung." This daily newspaper with a print run of nearly 200,000 targets the general public, and is also a regional newspaper of the capital city of Berlin. The fourth source we included in our analysis is the "Deutsches Ärzteblatt," which is a weekly journal with a print run of about 370,000 that primarily caters to physicians.

Apart from that, publications of medical organizations, self-help-groups, and the bpb[15] (center for political education) have been included in our analysis. Thus, a wide range of social representations is taken into consideration, as all important social systems will be included: science, medicine, economy, politics, self-help groups, and church. Of course, some information is directly derived from a social system (e.g., position papers published by self-help groups), other information is mediated by the media (e.g., reports about self-help groups in DIE ZEIT). The reference time period is from January 1st, 2002 to March 31st, 2004. As we focus on the situation in Germany, only articles published in Germany will be consulted.

Eighty-nine articles in DIE ZEIT, 8 in the Süddeutsche Zeitung, 59 in the Deutsches Ärzteblatt, and 5 in the Berliner Zeitung make reference to PGD.

[15] bpb stands for "Bundeszentrale für Politische Bildung."

A superficial first analysis of the material already reveals two interesting aspects: first, the number of articles on the topic in each newspaper or journal shows that most articles are printed in media that is highly academic or specialized (such as "DIE ZEIT" and "Deutsches Ärzteblatt"). In contrast to this, the number of articles in the Süddeutsche Zeitung, a widely read paper, is not high. Second, several focal points can be found: some articles report survey results (e.g., TAB 2002, Brähler and Stöbel-Richter 2004, Richter 2004 or Chrismon 2002). Other publications cover regulations in other European countries (Niew *et al.* 2002; Bahnsen 2002; Spiewak 2002; bpd 2004; Simm 2003; Süddeutsche Zeitung 2003). Another topic is telling about cases of families who had used PGD or who would like to resort to it (e.g., Spiewak 2002). Furthermore, many reports deal with German ethics councils[16] (Schüle 2002) or interviews with their members (Süddeutsche Zeitung 2003). Some of the articles discuss the ethical problems of PGD (German Association of Female Doctors 2004).

When turning to the discourse analysis, we will first present seemingly "objective" data. Information produced by the social system of science mostly just describes and explains the empirical results of surveys, relating what answers people have to given questions. Somehow, such data are assumed to be "objective," also because they are usually not combined with the author's personal views. After we have examined some examples of such studies, we will turn to material that more clearly reveals arguments for or against the legalization of PGD.

Seemingly "Objective" Data

First of all, we are going to draw attention to the social system of science, and to concentrate on publications that are of special interest with regard to the co-construction of society and technology. Thus, publications that "only" describe a technological procedure or notes about newly found genes will not be included. Within our context, it is more relevant to have a look at studies that reveal attitudes towards technology.

In general, several studies have revealed that German society is often ambivalent about technology. For many years, TAB, the Bureau for

[16]The National Ethics Council was inaugurated on June 8th, 2001 as a national forum for dialogue on ethical issues in the life sciences. It is intended to be the central organ for interdisciplinary discourse between the natural sciences, medicine, theology and philosophy, and the social and legal sciences, and to express views on ethical issues relating to new developments in the field of the life sciences and on their consequences for the individual and society (http://www.ethikrat.org).

Technology Assessment of the German Bundestag, has surveyed attitudes towards different fields of technology. Seemingly objective data is presented: although only 10% claim to be against technological progress, there are differences depending on the field of technology. Domestic or medical technological devices are accepted, whereas "risky" technologies, such as nuclear power or gene technology, are not (TAB 2002). The latter finding is of special interest as gene technology is also used for medical purposes. When people were asked about their opinion concerning PGD, TAB reported a very ambivalent picture, without mentioning details or percentages (TAB 2002).

Although TAB claims to be to be an "autonomous scientific institution" (http://www.tab.fzk.de), it seems to be quite closely linked to politics. This might be one reason why the authors of this study constantly repeated that people have become less sceptical and more accepting of technology. Technology and attitudes towards technology are presented in a way that reflects the values and goals of the social system that issues the information.

Another example for a scientific study is the most recent survey by Brähler and Stöbel-Richter (2004). This study, which is based on a sample of 2,000 Germans, demonstrates that most people have only a very poor knowledge of PGD: 30% of the respondents had heard about this technology at least once, while 60% had never heard of it at all. The main source of information was television (51.5%). People who knew something about PGD ranked their knowledge as fair or bad, rather than good. Most people overestimated the potential of PGD.

Nearly 80% thought that PGD can reveal whether the child would have severe physical or mental retardation, 53% believed that PGD can disclose everything, 23% reckoned that the child's body height, eye, and hair color could be predicted, and 16% thought that PGD can also tell what moral character the child will have. People were also asked the reasons for which PGD should be legalized and respectively whether they would like to resort to PGD themselves in the scenarios in Table 4.

People are more willing to support legislation of this procedure than they are willing to use it themselves. The difference is smallest for screening for intelligence, obesity and sex. This means that people have very clear views on these questions.

All in all, 61% would support the legalization of PGD if a certain disease were suspected. 31% would approve a general genetic check, 4% would favour PGD for the selection of other than health-related attributes, and 24% are against PGD.

Table 4. Motives for Using PGD

PGD should be legalized for ...	Would favour legalization (%)	Would like to use PGD themselves (%)
... severe diseases that lead to death by the age of one	76	63
... Down syndrome	70	60
... a disease that leads to death in young adulthood	62	50
... cancer, which has a high risk of onset in adulthood	40	32
... below-average intelligence	14	12
... risk of severe obesity	12	10
... the child being the "wrong" sex	8	6

In addition, people were asked about their thoughts and emotions when they think about PGD: 45% feel ambivalence, 44% uncertainty, 31% hope, 29% fear, 22% confidence, 17% arrogance, 13% consternation, 13% lack of comprehension, 9% paternalism, and 6% happiness.

These data confirm that people regard PGD with ambivalence and uncertainty. Positive and negative emotions are balanced. The study demonstrates that people do not know much about this technology, yet most of them state that they are willing to use it, while its potential is widely overestimated. Why PGD is socially represented in this way is not explained. The authors do not evaluate their results, nor do they draw any conclusions.

One should bear in mind that this study is based on a representative poll, and it is safe to assume that most people surveyed here would not have had any contact with this technology. A comparison with other studies, in which people who were somehow personally affected by this issue were included (couples with infertility problems or couples who are at risk of transmitting a genetic disease onto their children) provides further interesting insights.

When people with infertility problems were asked, the percentage of those wishing for PGD to become legal was much higher than in the survey by Brähler and Stöbel-Richter (2004): 97% of patients attending a reproduction clinic want PGD to become legal and 88% would like to use this technology themselves (Borkenhagen 2004). The author concludes that German couples with infertility problems will seek help abroad if PGD is not legalized in Germany.

A team of physicians (Krones *et al.* 2002) carried out a study in which different university hospitals collaborated. The project was part of the German Human Genome Project, which includes a small research field of ethics. The authors compared couples (n = 162) who are at high risk of transmitting a genetic illness to their children and couples with no known risk (n = 149). Eleven percent of couples at risk and 27% of couples without a known risk think that PGD should be kept illegal; 89% of the first group and 73% of the second group would be in favour of legalization. About 20% of couples at risk would like to go abroad to use PGD. As stated by Richter (2004), a professor for medicine, nearly 79% regarded the embryo that was ready to be implanted as "their child," and the other 21% described it as just a "cell cluster." In this study, the advantages and disadvantages of PGD were also listed. However, they were not a result of the study, but were compiled by the authors in advance. This might be one explanation for why they were formulated in a very technical manner.

The authors, all of whom were working at hospitals that could profit from legalization, seemed to be in favour of legalization. Evidence can be seen in the way the (dis)advantages of PGD were presented:

> Most probably, an abortion can be avoided and couples need not cope with the physical and psychological burdens associated with it. At the time of PGD, the embryo is in such an early stage that it can practically be ruled out that it has any feelings. (...) Children born after PGD seem to have no major health problems later on. By means of PGD, medicine could make great progress in the analysis and healing of inherited diseases. (Krones *et al.* 2002, translated by the author.)

The disadvantages listed here are mainly medical or technical difficulties; social and ethical problems are mentioned briefly (disposal of embryos, social pressure to use PGD, discrimination against handicapped people, and the risk of a new form of eugenics). Finally, the authors pleaded for "recognition of the views of people at risk" — 89% of whom want legalization.

Another study commissioned by a Christian organization asked for which purposes parents would like to use gene technology to improve their child's health (Chrismon 2002). Forty-two percent were of the opinion that man should not interfere at all. Fifty-five percent would like to use gene technology to prevent illness or retardation, 18% to improve intelligence, 16% to prolong expectancy of life, 10% to ameliorate the child's character and 6% to beautify the child. The authors, who carried out the study published by Chrismon, do not draw any conclusions but cite different people: Winnacker, a geneticist who is president of the DFG (German Research Foundation), who argued for the abandonment of children if parents cannot bear to have a disabled child; Peters, an actor, who is against any manipulation; and Westerwelle, the leader of the liberal party FPD, who is in favor of the legalization of any genetic manipulation that helps to cure severe inherited diseases.

The abovementioned studies are mostly presented as "objective data." Nevertheless, evaluation plays a role here, as one can tell from the authors' choice of study sample, the questions asked, and the way the results and conclusions have been presented. Further, it is of importance to know who paid for the survey.

Arguments in Favor of the Legalization of PGD

Some arguments in favor of legalizing PGD are more on a macro level, meaning that these aspects are relevant more or less for society as a whole, whereas aspects on the micro level mainly affect individuals, couples, or women, who come into contact with PGD. Although we will stick to this differentiation, some overlapping will persist. For example, general considerations about technology are attributed to the macro level, although they have effects on individuals at the same time. (The following passages present which arguments can be extracted from the mentioned documents. They do not represent the author's opinion.)

Arguments on the Macro Level

"No Need for Discussion, Just Imply PGD"

The core argument of this statement is as follows: history has highlighted how Germans have often been concerned about new technological developments, but in retrospect it has become obvious that there would have been nothing to worry about. Experiences of other countries show that there is no need to worry too much about this technology (e.g., FDP, liberal party). Ganten, a biotechnologist and head of the centre for molecular medicine claimed that PGD should just be allowed and put into practice; society will then learn to accept it (Michel 2003a). Ten years ago, there was also an intense discussion about IVF; now it is accepted. The same would happen with PGD. Furthermore, some authors stated that PGD is nothing but an early PD, which is legal and is generally practised; thus PGD should also be legalized (Gyn 2002).

"No Health Problems, but Many Opportunities"

Until now, no data hint at any health problems for children born after PGD (Krones *et al.* 2002). In contrast, using PGD can be an advantage for the parents, as PGD could enhance the success of IVF if only embryos that have passed a quality test are implanted (critically cited by Hüppe 2004).[17] The strongest and most indefeasible argument promises "great advances in the analysis and curing of genetic diseases." (Krones *et al.* 2002, translated by the author.) This statement easily brushes aside any counterarguments.

"Only a Small Number of People Will Use PGD; There is No Danger of Uncontrolled Selection"

One argument in favour of PGD is said to be the small number of people who might use PGD if it were allowed in Germany. Not so many couples would be expected to resort to this technology (critically cited by Hüppe 2004). One reason for this there are few forms of retardation that are genetically caused. Therefore, the selective potential of PGD is not big, as stated by van den Daele (2002), who is a sociologist and a member of the German National Ethics Council. According to Nippert, a professor for women's health, P(G)D will not bring along uncontrolled selection as feared by many antagonists (bpb 2004).

"PGD Will Have No Negative Effect on the Acceptance and Integration of Disabled People"

Likewise, some authors assert that society will not face any negative effects from the legalization of PGD, but would profit from it. For example, PGD will not change society or have an effect on the acceptance and integration of disabled people (van den Daele 2002):

> Society won't give retarded people notification of the termination of solidarity, if it allows a few parents to select embryos. (Translated by the author.)

Van den Daele also cited several statistics that confirm strong acceptance for disabled people; e.g., 90% think it is best to care for a child who has Down syndrome at home, 90% would not mind having neighbors with retardation, and 80% would not mind staying in a hotel where disabled people were also staying. The author concluded that the selection of embryos and discrimination after birth are not correlated phenomena (van den Daele 2002). Thus, he

[17] Hüppe is vice president of an ethics council, representative for the interests of disabled people, and member of the conservative party CDU.

concludes that prenatal selection will prevail, and that solidarity with disabled people will further increase. Other authors have even emphasized that couples who would like to use PGD express more positive attitudes towards disabled people than the average person (Richter 2004).

"German Society Will Lose a Lot If PGD is Not Legalized"

At the same time, some experts, such Ganten have argued that German society will forfeit a lot if regulations are not changed, as industry will lose economic connectivity and scientists with high potential specializing in PGD will leave the country and work abroad (Michel 2003a). Finally, legalization in Germany would put an end to "PGD-tourism" to neighboring countries (Borkenhagen 2004).

Arguments on the Micro Level

"PGD Strengthens the Individual's Autonomy"

One main advantage of PGD that has been cited is that PGD is helpful to people and strengthens their autonomy. Couples or women have the right to make an autonomous decision about their family planning (Bahnsen 2002). This opinion is also shared by some members of the German national ethics committee (Zypris 2003). As couples are sometimes said to have the right to have healthy children, PGD increases their autonomy by enabling them to have children of their own. Moreover, some say that parents do care about the health of their children and would use PGD only out of concern for their children's health (Gyn 2002).

PGD can help those who wish to have a child but who risk transmitting genetic mutations onto their children, or those who have infertility problems (critically cited by the German Association of Female Doctors 2004). PGD would enable those couples to have a healthy child of their own (e.g., Krones *et al.* 2002, Gyn 2002).

"PGD Eases the Burden"

Another frequently cited argument is that PGD could prevent abortions and the physical and psychological burdens that come with them (Zypris 2003 or Krones *et al.* 2002). Therefore, PGD is said to be less stressful for women than PD (critically cited by the German Association of Female Doctors 2004).

Counterarguments Against Legalization

In a next step, typical counterarguments found in the analyzed documents will be presented, again differentiating the macro and the micro level. Again, the presented arguments are the result of an analysis of the documents, and do not reflect the author's opinion.

Arguments on the Macro Level

"Just Because Other European Countries Practice PGD, this Does Not Mean That Germany Needs to Follow"

Just because PGD exists and is put into practice in other countries, this does not mean that PGD automatically needs to be legalized in Germany as well. Otherwise, everything that has ever been invented would be put into practice, and ethical discussions would no longer be needed. The same argument applies when confronted with the argument of "PGD-tourism." Having PGD remain illegal does not imply a disadvantage for the economy or the sciences, but could encourage the investigation of alternatives to control disease (German Association of Female Doctors 2004).

"Technology: Low Success Rate, Risk of False Diagnosis, and Negative Health Effects"

One disadvantage of PGD lies in the low success rate of the whole procedure. In the beginning, women have to take on the burden of IVF: the side effects of certain medications, a high rate of multiple pregnancies, pre-term births and mortality, as well as the risks of anaesthesia and medical intervention (Krones *et al.* 2002, German Association of Female Doctors 2004). When so many oocytes are needed, women might be reduced to a production unit for research activities — as argued by Loske, who is member of the Greens (Michel 2003b).

Because only one or two cell(s) are analyzed, there is a risk of a false or ambiguous diagnosis (German Association of Female Doctors 2004). Further, only a defined spectrum of genetic mutations can be diagnosed. Hence, PGD and PD cannot guarantee the birth of a healthy child because the child could have diseases other than the ones analyzed (Krones *et al.* 2002).

As the whole PGD procedure is quite new, studies on its long-term effects on both the child (Krones *et al.* 2002, Hüppe 2004) and the woman do not as yet exist (German Association of Female Doctors 2004).

"There are Many Problems in Differentiating between 'Normal' and 'Not Normal'"

Thus, one central question and possible counterargument is who will decide on what basis tests can or cannot be conducted (DIE ZEIT 2002)? How can limits be established (Gyn 2002)? And who will decide which result gives the embryo the right to live or not (Bahnsen 2002)? What will happen in the case of heterozygote diseases? In this case, someone is a carrier for an autosomal recessive disease (for example, for cystic fibrosis or phenylketonuria PKU). If and only if the person mates with someone who is also a carrier, will there be a 25% risk that their offspring will develop the disease in question. If such a carrier status is diagnosed in the pre-implanted cell, it could be expected that the embryo will be destroyed (German Association of Female Doctors 2004).

"Limits Will Be Broken"

Another counterargument states that the limits set on PGD will be transgressed (German Association of Female Doctors 2004). According to Hüppe (2004) the same experiences will occur as with PD. First, PD was established for special cases only, but the indications have been widened steadily over the last years. Whereas PD was intended to be offered to women over the age of 35, there are no longer any age restrictions. Also, PD was meant to be offered in cases of couples with genetic risks. Nowadays, 8% of all PDs are desired because women claim to be frightened of giving birth to a retarded child, although there are no known risks at all. Some authors refer to the experiences of other European countries where PGD is allowed. In Belgium, for example, indications for PGD have been extended (Spiewak 2002).

"People's Welfare is Only a False Pretence"

The welfare of childless parents might only be pretence. The pharmaceutical industry has a lot of power (Hüppe 2004). Hence, PGD is fostered for economic reasons and not for the good of the people. In general, there is not enough discussion about other interest groups such as industry or about the role of insurance companies (German Association of Female Doctors 2004).

"PGD Technology is Based on Selection"

Krones and others (2002) stated that the selection and disposal of embryos could be an ethical problem. But they did not explain the exact nature of the problem. Others have explained that the first selection is made in the laboratory. Then, experts — and not parents in the first place — choose which embryo to implant (German Association of Female Doctors 2004).

"Risk of Eugenics and the Economization of Life"

Several authors have discussed the risk of eugenic tendencies; whereas some have only mentioned them (Krones *et al.* 2002, Gyn 2002), others have warned against them (Hüppe 2004, Zypries 2003), and some have described them (e.g., van den Daele 2002). Some have reminded their audience of experiences with the Nazi regime in Germany, which should be enough of a warning not to allow PGD, such as Böckenförde (2003), a former judge at the Federal Constitutional Court.

PGD might enhance the view that health and fitness will become more important in society. Disabled people will be discriminated and stigmatized (Krones *et al.* 2002), and might experience great problems knowing they have a genetic anomaly that is defined as a reason for destroying embryos (Zypris 2003). In general, society could have problems accepting people with retardation (Rabbata 2003).

If the selecting of embryos becomes routine, society will lose its diversity (Simitis 2003). The legalization of PGD will entail research for which human embryos will be produced; thus, reproductive cloning and germ line manipulation might follow as well (German Association of Female Doctors 2004). According to Loske, member of the Greens, this development could also lead to an urge to use "surplus embryos," which would otherwise be destroyed (Michel 2003b or see e.g., Hüppe 2004), or, for example, for stem cell research (Rabbata 2003) and cloning (Böckenförde 2004). In general, the legalization of PGD could also open the door for designing babies (Gyn 2002, Simm 2003).

"Humans are Not God, and the Embryo Needs Protection"

Man should not play God (Berg 2003).

> Even the attempt (...) to use PGD in order to separate good and bad embryos would mean that mankind presumes to be the Creator, but mankind is not a match for that. Respecting the inviolability and sanctity of life involves understanding the narrowness of human rationality. (Jessen 2003, translated by the author.)

On that account, the human being should be granted human dignity — without any limits or expectations. Otherwise, we would be confronted with Peter Singer's discussion of the consciousness of self and rationality (Hüppe 2004).

Arguments on the Micro Level

"PGD Fosters Unrealistic Hopes"

The claim that PGD promises a better life is an unrealistic promise (Michel 2003b). Disease and suffering are part of human life and cannot be avoided by PGD (bpb 2004). PGD could raise false hopes that a child will be healthy — and if this is not the case, the disappointment could be extremely high (Krones *et al.* 2002).

"PGD Will Not Stop Abortions and Will Not Ease Psychological Burdens"

The argument that PGD prevents abortions cannot be proven, because PGD is mostly followed by PD (German Association of Female Doctors 2004; Hüppe 2004). There is no data that provides evidence of the psychosocial advantages of using PGD for women.

In general, some authors deny that there is such a thing as a right to healthy children (Gyn 2002). If a child is born after PGD this could result in psychological strain as well. Knowing that the child is the product of a technological procedure can be difficult for the parents and the child to deal with (Krones *et al.* 2002).

All in all, autonomy is not strengthened. Women might be forced to do everything in order not to give birth to a child that is not genetically "normal," and, in general social pressure to use this technology could increase (Krones *et al.* 2002, Zypris 2003, German Association of Female Doctors 2004).

"The Embryo Will Be Subject to the Parents' Wishes"

The embryo might become the subject of the parents' wishes and if it does not meet these expectations, it might be destroyed. PGD does not prevent disease — it only prevents (as far as possible) life with a child who has a genetic disease (van den Daele 2002). Parents could even select the embryo to be implanted based on sex (which is already happening in some countries) or eye color (Michel 2003b). If a couple is not willing to accept having a child who has an illness or who suffers from retardation, they should refrain from becoming parents (Böckenförde 2004).

Summary: Discourse on PGD in German Society

The discourse on PGD is often strongly connected with values and emotions. The analysis of the discourse about this technology shows that nearly each argument in favor of the technology has a counterargument. This is illustrated in Table 5.

Often, we can find social representations that are typical for certain social systems. For example, conservative statements are typical for conservative parties, liberal arguments for liberal parties, etc. Yet, in the case of PGD, arguments for and against PGD and can be found among members of all political parties. One exception might be the Greens, where most members appear to be against the legalization of PGD. However, not many clear statements on this issue have been published by the Greens.

Also, members of self-help groups for people with retardation are said to be torn between hopes and fears (Graumann 2002). Yet an analysis of official statements predominantly shows rejection of PGD (Bundesvereinigung Lebenshilfe 2003 or the Landesverband Bayern für Körper- und Mehrfachbehinderte (LVKM) 2003).[18] Instead, proponents of PGD can be found in chat or thematic groups (e.g., http://www.bundestag.de/dialog/forum/enquete_medizin_ archiv/02 lensingwerner.html).

Moreover, some other social institutions, like the DFG (German Research Foundation) have not published any comments that clearly express their position on PGD.

On the other hand, while some authors have hinted that physicians in other European countries hold a quite liberal view (Graumann 2002), medical associations in Germany have been among the most vocal opponents of PDG (cf. e.g., Rabbata 2003, or the German Association of Female Doctors 2004).

Although the discussion about arguments for and against PDG continues, there seems to be a development *towards the legalization* of PGD. For example, the German National Ethics Council (2003) concluded in its last statement of 2003 that it would support the legalization of PGD. Although the organization does not have any legislative power, politicians refer to its

[18] Both institutions are organizations for disabled people.

Table 5. PGD: Pros and Cons

Pros	Cons
History demonstrates that German society is always too anxious about new technologies.	Nazi history should be warning enough.
PGD is practised in other European countries.	This does not justify legalization in Germany. Some of the other European countries have even reported negative experiences.
PGD is nothing but an early PD.	The legal status of PD does not legitimate the legalization of PGD.
An embryo at this stage does not feel anything.	There is no exact knowledge of this. The embryo should be protected without any exceptions.
No health problems in children born after PGD have been registered.	PGD is too new to provide information about its long-term effects. It can be a burden for children (and their parents) to know that they are a technological product. IVF has serious negative effects on women's health; there are not enough studies about its long-term effects. Women might be reduced to being a production unit of oocytes.
PGD enhances the success rate of IVF if only examined embryos are implanted.	This is a dangerous form of selection and leads to the disposal or utilization for research purposes of embryos that have not been implanted.
PGD helps to prevent abortions.	PGD does not prevent abortions because people normally also use PD.
PGD can bring about great advances in the analysis and curing of inherited diseases.	PGD promises too much; there is even the risk of a false diagnosis, and only a defined spectrum of diseases can be diagnosed.
Only a small number of people will use PGD.	Indications for PD have also been widened over the time, so the number of people wishing PGD will be higher than first expected.
Only a small number of inherited diseases can be diagnosed, so there is no risk of selection.	The number will increase and people could request selecting for sex, eye color, or hair color. It is difficult to differentiate between "normal"/healthy and "not normal"/ill. There is a risk of eugenics.
The utilization of PGD and the acceptance of disabled people have nothing to do with each other.	PGD will cause discrimination and stigmatization of disabled people.
PGD is necessary to keep economic connectivity and to allow scientists with high potential to work in Germany.	A ban on PGD could spur new research on alternatives. PGD aggravates the economization of life.

Table 5. (*Continued*)

Pros	Cons
PGD strengthens people's autonomy.	PGD decreases autonomy and forces women to resort to this technology; it also reduces the autonomy of people affected by certain conditions.
People have a right to healthy children.	There is no right to have a healthy child, PGD does not guarantee health of the child.
PGD is a means of caring for the health of children.	PGD does not extinguish illness, but disposes of embryos with a certain genetic status.
PGD eases psychological burdens.	This has not been proven; the argument about people's welfare could be a false pretence. People are not God.

statements:

> Precisely in the field of personal decisions about the conduct of one's individual life, the limited legal approval of PGD would be most likely to achieve a just balance between the individual claim to freedom on the one hand and the protection of general legal goods by the State on the other. After all, a limited approval of this kind would enable couples wishing to have PGD to exercise responsibly their fundamental right to reproductive freedom without the implication that the moral position of those who strictly reject PGD was thereby devalued or declared untenable (2003).

It is interesting to examine the final decision recommending the restricted utilization of PGD: 15 council members voted for the legalization of PGD, and 7 against it. In both groups there were lawyers (most of them voted for legalization), and all of the physicians in the groups were in favour of legalization, while all church members were against it. The only representative of a self-help group (for Huntington Disease) was against legalization. The fact that all physicians voted *for* legalization is especially remarkable when the negative attitude of German physicians' associations is considered.

With reference to the work of the German National Ethics Council, politicians such as Bulmahn (former minister for education and research, social democratic party SPD), Ulla Schmidt (minister for health and social security, SPD party), and Renate Schmidt (former minister for families, seniors, and youth, SPD party) have expressed their appreciation of the final statement and have added:

> We appreciate very much the analysis of and opinion about PGD presented by the German National Ethics Council. When considering the interests of

affected women in particular, we think that a limited utilization is acceptable.[19] (Translated by the author.)

Reflecting on the discourse analysis itself, we should note that most of the discourse seems to be based on the opinions of experts. Either the experts themselves write about PGD or journalists report their opinions. Contrary to common assumptions, a broad public discussion (Graumann 2002) has not taken place, or is at least not observable in the documents that we referred to for this paper. This fact is reflected by the result that only a very few articles have been published in the Süddeutsche Zeitung, for example. We can conclude that the final decision about the legalization of PGD will represent more how PGD is socially constructed within the group of experts than within the general public. The supposed attitudes of the public are only referred to when they can be used to support an expert's argument.

Technological Development and Society: What Can Be Learned from the Discourse about PGD in Germany

If we turn back to our starting idea that technology can only develop within a social system, it seems that society has changed because the resolve *not* to allow PGD in Germany has weakened.

If we compare the social background of the time when the law for the protection of embryos was passed and the fact that this position was quite stable on the one hand, and social circumstances nowadays on the other, some important differences can be noted: when the aforesaid law was passed, society was a lot more critical of genetic technologies. The technology of PGD was not in an advanced state and its promises were quite vague. The economic situation in Germany was much better than it is now. Since then, many changes have taken place in Germany and abroad. Many countries have gained experience with PGD, and scientists have become famous because of their research results in this field. Different applications of genetic technologies have become evident, and more and more promises of future applications have been expressed. Germany perceives itself to be confronted with several problems, in particular economic and scientific weakness. This is reflected in the abovementioned discussion on whether German science and economy will lose out if PGD remains illegal. It is argued that the German education and science systems are in danger of losing their international connectivity. At the

[19] See http://www.bmgs.bund.de/deu/gra/themen/sicherheit/index_2906.cfm.

same time, there has been more individualization of risk and greater familiarization with other technologies such as prenatal genetic tests. Individuals have expressed concerns about possible problems with eugenics arising from the legalization of PGD, but at the same time would themselves like to use this technology. Furthermore, most individuals show excessively high expectations when asked about the potential of this technology.

Thus, the social context in which the discussion over the potential need to regulate technologies has changed. This explains to some extent why the discourse about PGD has changed, and why PGD is now socially represented in a different way.

As shown above, there is still no unanimity on the issue. Various reference anchors are used for the described construction of social representations. Yet, there is a noticeable tendency to emphasize the potential and positive side-effects of PGD. We should also bear in mind which social systems are significantly involved in the discourse about this technology. On the side of the supporters we find company representatives, scientists, and politicians — and thus, quite powerful systems. On the side of the opponents are a few professional groups, mainly physicians or self-help groups, and few scientists. Interestingly, the representatives of physicians in the abovementioned ethics committee are pro-legalization. Perhaps physicians experience pressure in their day-to-day practice because much of the responsibility for making decisions is placed upon them, such as in the case of abortions (in cases of medical indications, the law does not clearly state the point at which physicians may support the termination of a pregnancy, or when a prenatal genetic diagnosis can be said to be an "acceptable" reason for an abortion). Physicians who do not a great deal of direct contact with clients but who are more politically engaged might have a different point of view.

The whole of society is *not* included in the discussion. This result leads to the assumption that the discourse is *quasi-democratic* and *quasi-critical*. It is *quasi-democratic* because it tries to picture a situation in which *the* public or at least all important groups are included, e.g., via the ethics council. Policy makers can then pretend that all groups can participate. Not the whole of society is engaged in the discussion (considering the opportunity to hold a referendum, this would be possible). The discussion could be *quasi-critical* because it *must* be critical, given Germany's history. This became obvious through the analysis of the documents. Almost all supporters of the legalization of PGD also presented negative arguments. They stressed that they know about these arguments. Their decision for PGD could then be perceived to be rational — a matter of knowing and weighing all of the pros and cons. At the same time, it is often argued that in the case of PGD everything is different from the eugenic methods of the past. Quasi-critical is the discussion because it is simply presumed that the German public holds critical attitudes towards technology. Yet, as was demonstrated, the general public does not have much knowledge about human genetics. Attempts to educate the public and to include people in the discussion are not common. This leads to a central question: Does technology develop within *a* society?

The analysis presented in this paper results in the supposition that it is not *the* society that decides whether PGD is allowed or forbidden but that there are relevant social subsystems that are of importance in this respect. Which are the relevant social systems in our example? Mainly, the *experts* that are involved in the discussion. They influence how PGD is socially constructed to a great extent.

It can be expected that future developments on the issue will depend on the direction of the discourse. Turning finally to Luhmann (1989), one could expect that *economy* will align with genetic technologies in general and that PGD in particular when they influence its code. The economy's code is the ability or inability to pay. As the technology in question is said to be of great importance, it is likely that the social system of the economy will try to use PGD and therefore argue for legalization. The code of the *legal system* is to differentiate between justice and injustice. Hence, it will try to determine whether the legalization of PGD will interfere with existing laws. At the moment, it still does. But as demonstrated, there is a discussion on whether it would be "more lawful" to allow PGD. The social system of *science* has the code of truth and untruth. As the opportunities relating to PGD are perceived to those that provide people with "objective" knowledge, legalization would not interfere with truth but help to support it. The last example of an involved social system is *politics*. Its code is on whether or not to hold office. Two things could happen: either politicians are of the opinion that voters are against the legalization of PGD, in which case they would try to ensure that the situation does not change; or they think that the voters support PGD, in which case the politicians will struggle for legalization. As the cited studies clearly state that most people would favor legalization — and would be particularly interested in using this technology themselves — this explains why most politicians seem to be veering towards changing the current situation. All in all, we can assume a co-construction of technology and society, yet how technology is socially represented within different social systems is important, as this is what constitutes *the* society.

References

Bahnsen, U. 2002. *Check im Eikern. DIE ZEIT* 13/2002.

Berg, L. 2003. *Die Stammzell-Debatte war für uns ein Gewinn. Landesbischöfin Margot Käßmann zur Bioethik. Berliner Zeitung*, May 28, 2003.

Berger, P. and T. Luckmann. 1967. *The Social Construction of Reality: A Treatise in Sociology of Knowledge*. Harmondsworth: Penguin.

Böckenförde, E.-W. 2004. "*Dasein um seiner selbst willen.*" *Deutsches Ärzteblatt* 19.

Borkenhagen, A. 2004. *Unfruchtbare Paare in Deutschland wollen PID. Deutsche Ärzte-zeitung* http://www.aerztezeitung.de/docs/2004/06/30/120a0405.asp? cat=/medizin/sex/kuenstliche_befruchtung.

Bpb. 2004. *Gute Gene, schlechte Gene?* www.bpb.de/veranstaltungen/CQQ6WJ,0,0, Gute_Gene_schlechte_Gene.html#art0.

Brähler, E. *et al.* 2004. *Die Einstellungen der Deutschen zur Reproduktionsmedizin und zur Präimplantationsdiagnostik*. Unterlagen zur Pressekon-ferenz am 29.01.2004, http://www.uni-leipzig.de/~medpsy/pdf/presse_ repromedizin.pdf

Bundesärztekammer. 2001. *Konflikte bei ärztlichen Entscheidungen — am Beispiel der Präimplantationsdiagnostik*. Drucksache I-2, http://www.bundesaerztekammer.de/30/Ethik/10FortGenetik/01_06.html.

Bundestag Enquete-Kommission Medizin Drucksache 14/9020.

Bundesvereinigung Lebenshilfe. 2003. *Position der Lebenshilfe zur Präimplantationsdiagnostik im Überblick*.www.lebenshilfe.de.

Cara. 1992. *Schwanger sein ein Risiko?* Bremen.

Chrismon. 2002. *Was würden Sie an Ihrem Kind verbessern wollen?* www.chrismon.de/ctexte/2002/4/4-umfrage.html.

Deutsches IVF-Register 2001. Jahrbuch 2001. Bad Segeberg.

DIE ZEIT. 2002. *Gentest am Embryo*, 20, 2002.

DIE ZEIT. 2003. *Nationaler Ethikrat: "Keine einzige zwingende Lösung."* 05, 2003.

ESHRE Preimplantation Genetic Diagnosis. 2002. data collection III (May 2001). Human Reproduction, 17(1): 233–246.

German Association of Female Doctors. 2004. *Stellungnahme des Deutschen Ärztinnenbundes zur Präimplantationsdiagnostik*. http://www.kas.de/db_files/ dokumente/ die_politische_meinung/ 7_dokument_dok.pdf_1062_1.pdf.

German National Ethics Council. 2003. http://www.ethikrat.org/_english/publications/Stn_PID_engl.pdf.

German Society of Gynaecology and Obstetrics (Deutsche Gesellschaft für Gynäkologie und Geburtshilfe). 2003. *Engere Grenzen für Spätabtreibungen. Deutsches Ärzteblatt*, 28/29.

Gianaroli, L. *et al.* 1999. "Advantages of Day 4 Embryo Transfer in Patients Undergoing Preimplantation Genetic Diagnosis of Aneuploidy." *Journal of Assisted Reproduction and Genetics*, 16(4):170–175.

Graumann, S. 2002. Die Rolle der Medien in der Debatte um die Bio-medizin. In *Kulturelle Aspekte der Biomedizin. Bioethik, Religion und Alltagsperspektive*, edited by S. Schicktanz, C. Tannert and P. Wiedemann. Frankfurt: Campus.

Griffin, D. K. *et al.* 1992. "Dual Fluorescent in situ Hybridization for Simultaneous Detection of X and Y Chromosme-specific Probes for the Sexing of Human Preimplantation Embryonic Nuclei." *Human Genetics* 89:18–22.

Gyn. 2002. *Präimplantationsdiagnostik Pro und Kontra*. http://www.gyn.de/kinderwunsch/pid.php3.

Handyside, A. H., E. H. Kontogianni, K. Hary, and R. M. L. Winston. 1990. "Pregnancies from Biopsied Human Preimplantation Embryos Sexed by Y-specific DNA Amplification." *Nature* 344:768–770.

Handyside, A. H., P. N. Scriven and C. M. Ogilvie. 1998. *The Future of Preimplantation Genetic Diagnosis. Human Reproduction* 13/Supp. 4: 249–255.

Harper, J. C. 1996." Preimplantation Diagnosis of Inherited Disease by Embryo Biopsy: An Update of the World Figures." *Journal of Assisted Reproduction and Genetics* 13(2):90–95.

Holzgreve, W., D. Gänshirt-Ahler, and P. Miney. 1995. "Invasive und Nicht-invasive Pränataldiagnostik und Therapie in Utero." In *Pränatale Diagnostik und Therapie*, edited by R. Becker *et al.* Stuttgart: Wiss. Verl.-Ges.

http://www.bmgs.bund.de/deu/gra/themen/sicherheit/index_2906.cfm.

http://www.bundestag.de/dialog/forum/enquete_medizin_archiv/02lensingwerner.html).

http://www.mta-verband.at/zytogenetikforum/theorie/pid.htm.

http://www.spdfraktion.de/cnt/rs/rs_dok/0,20574,00.htm.

Hüppe, H. 2004. *Rede zur Präimplantationsdianostik*. http://www.aerzteblatt.de/download/files/X0001324.pdf.

Jessen, J. 2003. *Ist uns ein Klonkindl geboren? DIE ZEIT* 02/2003.

Keller, R. *et al.* 2001. "Zur Aktualität Sozialwissenschaftlicher Diskursanalyse — eine Einführung." In *Handbuch Sozialwissenschaftliche Diskursanalyse*, edited by R. Keller, A. Hirseland and W. Schneider. Opladen: Leske + Budrich.

Krones, Z. K. *et al.* 2002. *Präimplantationsdiagnostik aus der Sicht von Hochrisikopaaren und der Bevölkerung in Deutschland*: Erste Ergebnisse einer multizentrischen Studie. http://www.dhgp.de/media/xpress/enomexpress03_02/preimplantation.html.

Landesverband Bayern für Körper- und Mehrfachbehinderte (LVKM). 2003. *Selektion im Reagenzglas*. PID stellt Existenz Behinderter Menschen Infrage. http://www.lvkm.de/pressemi_200203.html.

Levinson, G. *et al.* 1995. "DNA-based X-enriched Sperm Separation as an Adjunct to Preimplantation Genetic Testing for the Prevention of X-linked Disease." *Human Reproduction* 10(4):979–982.

Luhmann, N. 1989. *Ecological Communication*. Chicago: The University of Chicago Press.

Michel, J. 2003a. *Blick in die Zukunft. Berliner Zeitung*, February 20, 2003.
Michel, J. 2003b. *Frauen dürfen nicht zum Rohstofflager werden.* Grünen-Experte Loske über das Klinen, die Ware Mensch und den Boom der Reproduktionsmedizin. Berliner Zeitung, February 20, 2003.
Moscovici, S. 2001. *Social Representations.* New York: New York University Press.
Munné, S., J. Grifo, J. Cohen, and H.-U. G. Weier. 1994. "Chromosome Abnormalities in Human Arrested Preimplantation Embryos." *American Journal of Human Genetics* 55:150–159.
Munné, S. *et al.* 2000. "Outcome of Preimplantation Genetic Diagnosis of Translocations." *Fertility and Sterility* 73:1209–1218.
Murken, J. and H. Greve. 1994. *Humangenetik.* Stuttgart: Enke.
Niew *et al.* 2002. *Bioethik. DIE ZEIT*, 05/2002.
Nippert, I. 2001. *Was kann man aus der Bisherigen Entwicklung der Pränataldiagnostik für die Entwicklung der Qualitätsstandards für die Einführung neuer Verfahren wie der Präimplantationsdiagnostik gelernt werden? Schriftenreihe des Bundesministeriums für Gesundheit. Fortpflanzungsmedizin in Deutschland* 132, edited by Bundesministerium für Gesundheit: 293–321.
Rabbata, S. 2003. *Konflikte programmiert. Deutsches Ärzteblatt* 1–2.
Richter, G. 2004. "Präimplantationsdiagnostik: Möglichkeit zur Erfüllung des Kinderwunsches." *Deutsches Ärzteblatt* 101(6).
Schüle, C. 2002. *Bioethik. DIE ZEIT* 31/2002.
Simitis, S. 2003. *Regeln für die Fortpflanzungsmedizin*, Interview mit Simitis, geführt von Ulrich Bahnsen und Andreas Sentker. *DIE ZEIT* 05/2003.
Simm, M. 2003. *Ungleiches Recht für alle. Süddeutsche Zeitung*, October 29, 2003.
Spiewak, M. 2002. *Schwanger um jeden Preis. DIE ZEIT* 20/2002.
Spiewak, M. 2003. *Mutterglück im Rentenalter. DIE ZEIT* 05/2003.
Staessen, C. *et al.* 1999. "Clinical Experience of Sex Determination by Fluorescent in situ Hybridization for Preimplantation Genetic Diagnosis." *Molecular Human Reproduction* 5:382–389.
Süddeutsche Zeitung 2003. *"Designer-Baby" soll den Bruder Retten.* June 20, 2003.
TAB 2002. *Technikakzeptanz und Kontroversen über Technik*: "Positive Veränderungen des Meinungsklimas — konstante Einstellungsmuster." Ergebnisse einer Repräsentativen Umfrage des TAB zur Einstellung der deutschen Bevölkerung zur Technik. http://www.tab.fzk.de/de/projekt/zusammenfassung/ab83.htm.
Van den Daele, W. 2002. *Zeugung auf Probe. DIE ZEIT* 41/2002.
Verlinsky, Y. *et al.* 1994. "Preimplantation Diagnosis of Genetic and Chromosomal Disorders." *Journal of Assisted Reproduction and Genetics* 11(5):236–243.
Verlinsky, Y. *et al.* 1997. "Current Status of Preimplantation Diagnosis." *Journal of Assisted Reproduction and Genetics* 14:72–75.
Zypris, B. 2003. *Vom Zeugen zum Erzeugen? DIE ZEIT* 45/2003.

Part 2
New Media and Generations

5 Analysis of the Content of Newsgroup Messages: Methodological and Technical Issues

Luca Giuliano

Introduction

Computer Mediated Communication (CMC) has reached remarkable dimensions in recent years. Although this fact is still too recent to have undergone analysis by academics through a regular framework (Castells 2000), CMC has contributed to the birth and making of "virtual communities." The peculiarity that such communities have in common is that they share a "digitalized territory": a sort of "cyberspace" not opposed to "true life" but interlaced with it in many ways (Giuliano 1999; 2002a; Paccagnella 2000; Baym 2002).

Cyberspace is one of the places of the media landscape that belongs nowadays to a large number of people in industrialized countries. On September 3rd, 2004, there were 943 million Internet users according to the Computer Industry Almanac (http://www.c-i-a.com/), of which 45% were from G8 countries and constituted 49% of their population. "Cyberspace citizens" are not only information consumers but have become information "producers" themselves in a multicultural environment, able to overcome language barriers. The dominance of the English language has not prevented the diffusion of other languages, while automatic translation programs (actually still very imprecise) are becoming widespread. The transformation of users into information producers is the peculiar characteristic of a collective intelligence that, according to De Kerckhove (1991) and Galimberti and Riva (1997), can be named the ***inter-brainframe***. The inter-brainframe is an innovative trait, compared to the television mental framework (***videoframe***) that generated our mass culture, an inborn feature of the mental cybernetic framework (***brainframe***).

The brainframe that distinguished the birth of the digital culture implies the co-presence of a large variety of cognitive frames on the net (1997).

This newly-born anthropological space sets the boundaries of a "knowledge space" according to Pierre Lévy (1994) (Day 2001), which consists not only of scientific and technical competence, but also of a "semiotic space" made up of emotions, learning, experiences sharing, and knowledge. The knowledge space in the Internet is mostly hidden, since it is not even reached by search engines. Thus, it stays out of any kind of possible reading, interpretation, or explanation, in spite of its importance for sociological analysis. In this space made up mostly of "words," one needs to have the tools to surf and orient oneself in the internet. On the Internet, anyone on the web may write and "publish" anything without any pre-set selection criteria. Extracting interesting information or at least finding meaningful kernels of information in digitalized space is (and will become even more so in the future) a real effort. To some extent, search engines help with "information retrieval" on the web, but the technology is incapable of carrying out any semantic analyses of the contents. The "Semantic web" that has been widely discussed recently, is a technology that inquires about texts that process metadata; i.e., an information framework on textual data, which is, however, at the initial stage.

In this report, we examine newsgroups, specifically those focusing on political issues. In general, the methodological and technical problems in analyzing the contents of these messages are, for the most, part identical to those encountered during the analysis of e-mail messages. Yet other aspects of newsgroups have peculiar characteristics that make them especially interesting for sociological analysis. Their importance for the observation of linguistic behavior and mass interaction as an environment has been widely documented (Resnick 1997; Paccagnella 1997; Stubbs 1998; Fredrick 1999; Pritchard 1999; Fiore *et al.* 2002; Rosen *et al.* 2003).

Newsgroups are "electronic notice boards," each focused on a different topic, where users can post their messages and read the messages of other users. Entry into a newsgroup requires a specific program (*newsreader*) and access to a place that can be defined as the "notice board room," i.e., the *news server.* Since 1995, the archiving of messages has become user-friendly through the DejaNews web interface. In February 2001, the entire archive (starting from 1985) was purchased by Google, which has made it available through its search engine.

Usenet newsgroups form a substantial portion of the asynchronous electronic messages that generate traffic of about 1,000,000 messages per day, equivalent to 70–80 terabytes per year (Lyman *et al.* 2003). Newsgroups scattered worldwide and classified according to their language group and topics number more than 25,000 units according to Choi and Danowski (2002), and 50,000 according to Smith (2002). The uncertainty over the figures is due to the steady growth of newsgroups and to the fact that not all servers provide access to the same kinds of newsgroups. A study (now out-of-date, but still significant) carried out on the whole corpus of Usenet messages (Smith 1999) throughout 150 days in 1997, counted 14,347 newsgroups, with traffic of 6 gigabytes per day. On average, 20,000 users sent 300,000 messages every day. More than 1,000,000 people sent at least one message during the period under study.

Newsgroups are remarkably interesting to those studying CMC and digital culture. The Usenet system is user-friendly and has been well known worldwide since 1979. In recent years, it has grown to include a complete range of topics branching out in many linguistic and cultural areas. The Usenet system involves millions of people, and copes with such a complexity of topics and social interactions that specific analytical methodologies are required (Whittaker *et al.* 1998).

Research Method and Goal

Our research goal is to explore the potentiality of the application of a text mining strategy and of the automatic analysis of textual data. The aim is to identify, with limited means, the general contents of the messages, the users' main themes and, possibly, their opinions. This is actually a very ambitious goal as the problems encountered in this kind of research are, generally speaking, quite serious. On the other hand, the textual data and text-mining analysis software programs useful for our purpose have greatly improved in recent years and are still being used. To mention a few examples: the papers of the last two JADT meetings — Journées internationales d'Analyse statistiques des Donnée Textuelles (JADT 2000, 2004) show regular progress concerning these issues.

The major difficulty in carrying out a content analysis of this material is the huge amount of data to be processed and interpreted that is bothered by "background noise," which is often difficult, if not impossible, to eliminate (Giuliano 2003). In the communication scheme by Shannon and Weaver, the "noise" represents an unintentional event that is interfering with the decoding of messages by the receiver. This interference can be displayed at a technical level (reception noise) or at a semantic level, because of the distortion of the message's meaning, given by the existing differences in cultural and linguistic codes between senders and receivers.

As regards our study "the background noise" is that portion of information (intentional or unintentional) that is useless to the understanding of text content, or better still, represents an element of disturbance because it hinders the correct identification of the existing kernels of meaning.

A well-known property of Usenet messages is that they have a low value in the ratio of signal/noise. This "value" decreases more and more when a greater number of users send uselessly long, redundant messages, unrelated to the theme, which openly breaks the rules of *netiquette* (Smith 2002).

Noise sources in *newsgroups* can be classified in three big categories: (1) environmental noise; (2) rhetorical noise; and (3) argumentative noise.

Environmental noise is related to the nature of an electronic message. The message can contain linguistic signs that are unidentified by the server and decoded incorrectly because different codes are used by the many operating systems. In a message, there are email addresses, web-links, and numerical sequences derived from codes existing in the body of a message that record the route, date, and further details of the message. Occasionally, messages may also record alphanumerical codes in their text, deriving from the codification of the attached images.

Rhetorical noise is given by the kind of message and by its communication strategies. It presents acronyms, quotations of messages that the user has replied to (*quoting*), jargon, recurrent signatures (*signatures*) reporting the author of the message, and brief commercial ads plugged automatically by free mail servers.

Argumentative noise is created by deliberate actions of disturbance made by provokers (*trolls*) that try to foment emotional and quarrelsome responses from the regular group users. "*Troll*" is a word meaning both a fishing technique and an ugly, stinking, and unpleasant fairy tale creature. A *troll*, if not ignored, may often stir up a "web fight" (*flame*). Quite often, these provokers navigate politics-focused newsgroups on purpose.

The methodology used is the automatic content analysis aimed at identifying the most adequate key words that describe the contents of the messages. The procedure for treating the corpus involves the following steps:

- Normalization of the texts and an analysis of the main lexicon-metric characteristics;
- Selection of a sub-corpus relevant to the concordances of the form "war";
- Mining of repeated segments;

- Making of lexical tables segments/newsgroups and graphic forms/concordances.
- Analysis of the correspondences on the lexical table and classification of the concordances.

This procedure inevitably implies a loss of information and approximation in semantic analysis. But it allows the problem of background noise to be successfully overcome. Although the *noise* created in newsgroup messages at a first scanning based on the reading of the messages is quite significant, it is eventually reduced when techniques of textual analysis and multivariate analysis are applied. The main graphic forms are selected according to strategies that bear in mind a frequency threshold, text coverage quantity (extent, size), and the distinguishing characteristics of the text. The software in use is TALTAC 1.6 (Bolasco *et al.* 2000–2003) and the two versions of SPAD-T, 1.5 per DOS and the version of the same program for Windows (Lebart *et al.* 1993).

Results of the Application to Newsgroups of the Hierarchy It-politica during the Iraqi War

Before describing the results of our study, we want to remind outsiders to this field of study that:

a. The condition for the drafting of the text of messages in a newsgroup is the result of an asynchronous interaction procedure, with an implicit tone of public discussion and mostly with an informal style (although there are messages deriving from much more formal form of communication, such as a newspaper article or a bulletin from a press agency). The interaction is voluntary, potentially global for its admittance procedure and anarchical in management. Very few newsgroups have an authorized moderator to screen the incoming messages.
b. Messages are made up of a header (*header*), which includes a group of lines of texts followed by the "body" of the message (*body*). Newsgroups writers use a specific lexicon typical of web communication, with emoticons (even if the so-called "faces" are today quite rare), jargon words, and acronyms.

The corpus IRAQ28 (Table 1) is made up of messages posted in newsgroups of the site it.politica during the Iraqi war, specifically in the month from March 18th to April 18th, 2003.

Table 1. Newsgroups of the Hierarchy It.politica: Messages Posted in the Pperiod: 18th March–18th April 2003

Newsgroups	N. messages	Average msgs in a day
it.politica	13,931	435.34
it politica.internazionale	6,194	193.56
it.politica.pds	2,981	136.25
it.politica.lega-nord	3,809	119.03
it.politica.polo	4,496	112.53
it.politica.rifondazione	2,728	90.91
it.politica.destra	686	21.44
it.politica.cattolici	667	20.84

Table 2. Lexicon-metric Measurements of Corpus IRAQ28

Occurrences	N	5,220,932
Graphic forms	V	179.112
Type/Token ratio	(V/N) * 100	3.43
Hapax percentage	(V1/V) * 100	46.85
General average frequency	N/V	29.49

The corpus is made up of 35,492 messages (33 Mb). Its text has been subjected to complete normalization with TALTAC 1.6, achieving the lexicon-metric measurements in Table 2.

The corpus, though stripped of the *header* of the messages (containing addresser and addressee, forwarding date, object, mail server, etc.), displays a high percentage of "noise" from different sources: HTML markers, Internet and email addresses, server unidentified characters, spelling mistakes, "signatures" of addressers using drawings in ASCII characters, emoticons, jargon words, or exclamations, quotations, the copying and pasting of web pages, the overlapping of different idioms, and so forth.

The large amount of noise included in the corpus can be approximately assessed through graphic forms not identified by the TALTAC grammatical tagging (see Table 3).

The unidentified graphic forms number 76,984 (equivalent to 42.98% of the distinct graphic forms versus an average of 5% in literary or press texts). In spite of the very high percentage of noise, this can be considered insignificant in the analysis of the lexicon and in a multidimensional text analysis. As a matter of fact, the major amount of unidentified forms (26.67% of the graphic forms) is situated in the hapax (graphic forms of occurrence 1). In fact, the main words of a language can be found especially in the first deciles

Table 3. Graphic Forms Not Identified in the Corpus IRAQ28

Graphic forms not identified	W	76,984	
Hapax forms not identified	W1	49,564	
Percentage of forms not identified	(W/V) * 100		42.98
Percentage of hapax forms not identified	(W1/V1) * 100		59.91
Percentage of hapax forms not identified out of the total amount of graphic forms	(W1/V) * 100		26.67

of the low frequency band and provide likewise the most useful information for techniques of multidimensional analysis (Bolasco 1999), whereas, unidentified forms tend to concentrate in the last deciles of the low frequency band. Assuming a frequency threshold of 17, recommended by TALTAC as reference for an automatic text analysis, the percentage of noise would drastically decrease to 5.04% of the occurrences, with a text coverage of 90.98%.

Another very important noise element is given both by the interventions of annoyers and by redundancy, caused by signatures quoting greetings or very frequent "mottos" that alter the lexicon of the actual "message body" (*body*).

Empirical studies dating back to the first studies on "virtual communities" (Rheingold 1993) have classified active and passive users. Active users are those who participate in the discussions. Passive users or *lurkers* (onlookers) are those who only read, without participating in the newsgroup. The number of passive users cannot be calculated. Messages can be read and downloaded without registering, so there might potentially be several thousand users. Specific investigations that have recently been carried out evaluate them at around 90% of users (Nonnecke and Preece 2000). We know nothing about such users. I the evaluation is correct, we have to consider that if a newsgroup is visited by an average of 500 users per month, in fact one can expect that there may be 5,000 others who are only "readers." Furthermore, there are huge differences in behaviour among the active users. Twenty-seven percent of users posted only one message during an observation term of 6 months (the average was 3.1 messages per user); while 2.9% of users posted 25% of all messages (Whittaker *et al.* 1998). Among the latter group are a very few "obsessive" users who post several dozen messages per day. This was also confirmed by our study.

In the <it.politica.lega-nord> newsgroup, we counted 586 active users during the period of the study. The first 50 users posted 66.06% of the messages. This distribution was confirmed in all eight newsgroups classified in the category <it.politica>. The total number of active users in the eight newsgroups was 2,390, but the first 50 users posted 43,27% of the messages.

Word strings (signatures, greetings, repeated sentences) creating disturbance and denoting subjects, are typically magnified in a multivariate analysis. Preliminary treatments usually reveal just these anomalies, which can be neutralized with a careful scan in the following analysis.

Corpora that have a large size, such as IRAQ28 (or greater), become effectively manageable by creating corpora that are reduced automatically, on the basis of themes identifiable through one or more key words. In our study, the word that most frequently appeared in the corpus (leaving out grammatical forms and trivia relating to Internet language) was "war" [*guerra*]. We built a reduced corpus of concordances based on that word, including 18,698 text strings, for a total of 474,476 occurrences and 32,945 distinct graphic forms. Thus, the maximum percentage of environmental and rhetorical noise has been reduced to 3.69%.

To examine the sub-corpus content we relied on identifying the issues that were being debated, through the frequency of semantic kernels displayed by the segments repeated in the concordances of the pivot form "war" (Table 4).

The observed newsgroups revealed a predominant opposition to the war. The most frequently repeated segment, leaving out grammatical forms, was <against the war>, with 780 occurrences. The newsgroups differed in their political orientation, with the exception of two general newsgroups ("politica" and "internazionale") that were devoted to Italian politics and international politics, respectively. The remaining ones reveal a clear ideological attitude: <it.politica.destra> sustains debates on ultra-right wing culture; <it.politica.lega-nord> deals with issues relevant to federalism; <it.politica.polo> is devoted to the majority parties belonging to the "Casa delle Libertà" party; <it.politica.pds> is devoted to the democrats of the Left

Table 4. Analysis of the Correspondences of the Segments Repeated Per Text: the Mining of Self Values

Self values	% inertia	% Cumulative inertia
0.0685	22.37	22.37
0.0587	19.17	41.54
0.0522	17.05	58.59
0.0415	13.57	72.16
0.0338	11.03	83.19
0.0287	9.36	92.55
0.0228	7.45	100.00

Table 5. Analysis of the Correspondences of the Segments Repeated Per Newsgroup: Coordinates and Absolute Contributions of Active Frequencies

Newsgroup	Coordinates		Absolute contributions	
	F1	F2	F1	F2
CATTOLICI	0.31	0.86	6.9	59.8
DESTRA	−0.49	−0.02	13.1	0.0
INTERNAZIONALE	−0.32	0.15	39.2	10.7
POLITICA	0.25	−0.04	26.6	0.7
LEGA-NORD	0.12	−0.19	1.7	4.4
PDS	−0.11	−0.29	2.1	16.9
POLO	0.29	−0.04	9.7	0.2
RIFONDAZIONE	−0.08	−0.23	0.8	7.3

party; and <it.politica.rifondazione> deals with issues concerning the far left wing linked to the Rifondazione Comunista.

In spite of these differing orientations, newsgroup users do not differ sharply in their net behavior. The practice of posting part of a message to all newsgroups simultaneously is widespread (*crossposting*), even though it is forbidden by *netiquette*. The users enjoy discussing their favorite topics with their "opponents." It often happens that some users indulge in provocation, pretending to belong to the opponents' political party or bombarding the newsgroup with insults.

An analysis was conducted of binary correspondences of segments/texts with a threshold of 20, performed with the SPAD 5.0 program that identifies thematic clusters peculiar to the different newsgroups.

Thereafter, we have examined the first two factorial axes that explain 41.54% of the total variance (Table 5). The coordinates and absolute contributions of the active frequencies allow an expected diversification of the first factor between the two main newsgroups <it.politica> and <it.politica.internazionale> to be identified, around the axis of foreign and domestic policy. The placing of the newsgroup <it.politica.destra> on the international side was expected, since in the past it had hosted a large number of messages denoting strong adherence to the principles of ultra right wing, anti-Zionist groups, with few of the messages showing concern for domestic policy issues.

The second factorial axis is characterized by the contraposition between <it.politica.cattolici> and <it.politica.pds> on the themes focusing on the debate over opposition to or support for the war and on the relevant motivations. Table 6 displays only the recurrent segments that contribute significantly to at least one of the two mined factors.

Table 6. Analysis of the Correspondences of the Repeated Segments with a Threshold of 20 Per Newsgroup: Coordinates and Absolute Contributions of Segments

SEGMENTS	SEGMENTS	Coordinates		Absolute contributions	
		F1	F2	F1	F2
A guerra finita	when war is over	−0.16	−0.35	0.1	0.8
amministrazione Bush	Bush administration	−0.41	0.13	1.3	0.2
Amnesty International	Amnesty International	−0.54	0.36	1.3	0.7
concetto di guerra	war concept	0.41	0.12	0.7	0.1
Consiglio di Sicurezza	Security Council	−0.62	0.15	2.6	0.2
contrario alla guerra	against the war	−0.47	0.07	3.1	0.1
contro gli americani	against Americans	0.08	−0.48	0.0	0.9
contro l'Iraq	against Iraq	−0.31	0.06	2.7	0.1
contro tutte le guerre	against all the wars	0.68	0.63	2.2	2.1
criminale di guerra	Criminal war	−0.59	−0.17	2.6	0.2
dichiarazione di guerra	war declaration	−0.56	0.62	1.4	2.1
diritto internazionale	international law	−0.41	0.15	1.7	0.3
favorevole alla guerra	Pro the war	0.42	0.15	1.1	0.2
fermare la guerra	stop the war	0.07	−0.47	0.1	6.3
grande guerra	big war	0.36	0.57	0.6	1.7
guerra breve	short war	0.05	−0.67	0.0	2.1
guerra dei trent'anni	Thirty year war	0.81	−0.17	2.3	0.1
guerra di aggressione	war of aggression	−0.44	−0.28	1.5	0.7
guerra di Bush	Bush's war	0.23	−0.15	0.7	0.4
guerra di difesa	defense war	0.49	0.89	0.7	2.8
guerra di liberazione	liberation war	0.22	0.14	0.8	0.3
guerra è guerra	war is war	0.62	−0.27	1.3	0.3
guerra finisca	war end	0.52	−0.26	1.0	0.3
guerra giusta	fair war	0.52	1.01	4.4	19.5
guerra illegale	illegal war	−0.43	0.02	1.0	0.0
guerra illegittima	illegitimate war	−0.10	−0.41	0.1	1.1
guerra in Afghanistan	war in Afghanistan	−0.39	0.18	1.0	0.3
guerra irachena	Iraqi war	−0.07	−0.56	0.0	3.2
guerra legale	legal war	0.94	−0.16	3.0	0.1
guerra lunga	long war	0.39	−0.45	1.7	2.8
guerra nucleare	nuclear war	−0.51	−0.01	1.0	0.0
guerra santa	holy war	0.26	0.17	0.8	0.4
guerra voluta	wanted war	0.18	−0.42	0.1	0.8
guerre giuste	fair wars	0.43	0.35	0.9	0.7
guerre mondiali	world wars	−0.15	−0.51	0.1	1.4
in caso di guerra	in case of war	−0.63	−0.07	1.5	0.0
legittima difesa	self-defence	0.78	0.06	2.2	0.0
liberazione dell'Iraq	Iraq liberation	0.68	0.09	1.6	0.0
movimento contro la guerra	movement against the war	−0.78	0.12	3.1	0.1

Table 6. (*Continued*)

SEGMENTS	SEGMENTS	Coordinates		Absolute contributions	
		F1	F2	F1	F2
Nazioni Unite	United Nations	−0.12	0.35	0.2	1.6
no alla guerra	no war	0.25	−0.02	2.5	0.0
nostro paese	our country	0.40	0.59	1.1	2.8
nuova guerra	new war	−0.48	0.07	1.6	0.0
opposizione alla guerra	opposition to war	−0.31	−0.25	0.8	0.6
popolazione civile	civil people	0.42	0.96	0.7	4.3
popolo iracheno	Iraqi people	0.38	−0.19	2.1	0.6
prigionieri di guerra	war prisoners	−0.31	0.18	1.6	0.6
prossima guerra	next war	−0.58	−0.02	1.1	0.0
quarta guerra mondiale	fourth world war	−0.13	−0.31	0.1	0.8
ripudia la guerra	repudiate war	−0.03	−0.35	0.0	1.8
scoppiata la guerra	war burst	0.47	−0.48	0.7	0.9
seconda guerra del golfo	second Gulf war	−0.28	−0.54	0.3	1.3
termine guerra	war ending	0.57	−0.20	1.1	0.2
Unione Europea	European Union	−0.25	0.05	0.7	0.0
uranio impoverito	depleted uranium	−0.61	−0.15	1.7	0.1
vittime civili	civil victims	0.57	−0.13	2.8	0.2

The contents emerge more clearly when we project recurrently mentioned segments on the factorial plain of axis 1 and 2 (Figure 1). In the first quadrant (++), where the newsgroup <it.politica.cattolici> is situated, we find segments representing the positions that are differentiated around the "humanitarian" theme of the war: <fair war>[<guerra giusta>], <against all the wars> [<contro tutte le guerre>], <civil people> [<popolazione civile>], <holy war> [<guerra santa>], <war concept> [<concetto di guerra>], <Iraq liberation> [<liberazione dell'Iraq>].

In the opposite quadrant (–) we find themes with a stronger political connotation displaying a sharp opposition to this <second gulf war> [<seconda guerra del golfo>]: <illegal war> [<guerra illegale>], <illegitimate war> [<guerra illegittima>], <war of aggression> [<guerra di aggressione>]. Inside this quadrant are the two newsgroups tied to the left-wing opposition: <it.politica.pds> and <it.politica.rifondazione>.

On the third quadrant (+ –), made up by the positive semi-axis of axis 1 and by the negative semi-axis of axis 2, we find the segments pointing to the motivations behind the war: <Bush's war> [<guerra

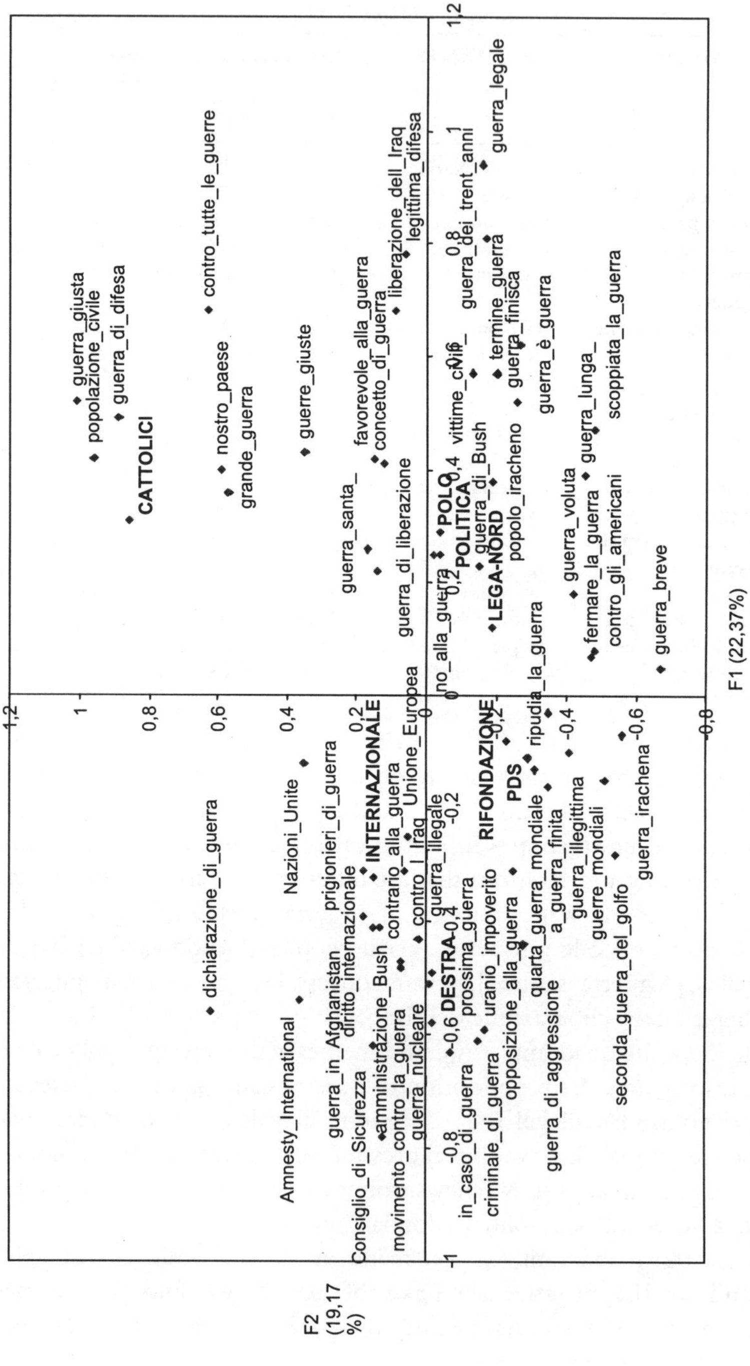

Figure 1. Projection of Repeated Segments on the Factorail Plane (Axes 1 and 2)

di Bush>], <war wanted> [<guerra voluta>], <against americans> [<contro gli americani>], <Iraqi people> [<popolo iracheno>], <civil victims> [<vittime civili>]. There are also references aimed at justifying military intervention: <legal war> [<guerra legale>], <war is war> [<guerra è guerra>]. In this case, too, there are segments denoting a debate for and against the war: <stop the war> [<fermare la guerra>], <war ends> [<guerra finisca>], <no war> [<no alla guerra>], <short war> [<guerra breve>], <long war> [<guerra lunga>]. The newsgroup <it.lega-nord> is mostly situated in this quadrant.

On the fourth quadrant (– +), made up by the negative semi-axis of axis 1 and by the positive semi-axis of the axis 2, there are segments referring to the international organizations (<United Nations> [<Nazioni Unite>], <Security Council> [<Consiglio di sicurezza>], <European Union> [<Unione Europea>], <Amnesty International>) and to the decision of war collocation in the context of international legitimacy, with clear references to positions against the war. The newsgroup <it.politica.internazionale> is strongly characterized by discussions on these issues.

Identification of the themes debated by newsgroups can also be performed through an analysis of the co-occurrences within the concordances of the pivot form <war> [<guerra>].

An analysis of the correspondences of the lexical table attaining to forms/concordances (*mots-réponses*) performed at the frequency threshold of 50, with the phase ASPAR of the program SPAD-T 1.5 for DOS, allowed the identification of two major factors that highlight the itemized meaning-kernels discussed in the examined period.

To obtain a more effective synthesis of this result, a classification of concordances has been worked out. The leading hypothesis of this analytical strategy is that by identifying the discussed themes, one can trace and quantify the weight, even though approximately, that each of them carries, within the messages in the period under observation. The weight that thematic groups carry is given by the comparative frequency of the concordances, classified within the respective groups. In the cluster analysis performed on the concordances, four thematic groups were identified.

As Table 7 shows, the main theme is n.3 (32.52% of concordances). The graphic forms classified within this group identify the topic of major interest for newsgroup users: the debate over the reasons for military intervention and its justifications from the point of view of international law. The peculiar forms in this group refer to Saddam Hussein's crimes, to the terrorist threat, to the

Table 7. Thematic Groups Resulting from a Cluster Analysis of the Concordances (N = 18,698) and Characteristic Forms for Each Group

Thematic group 1 — References to Italian foreign policy and to the controversy between the majority parties and the opposition (29.88% of the concordances).

UNO, fair, Pope, to exist, left, European Union, position, France, resolution, Italy, opposition, Germany, preventive, participate, God, legitimate, doctrine, international, moral, right, words, favorable, political, legal, politics, duty, D'Alema, against, word, name, Fassino, illegal, church, Berlusconi, to prevent, place, power, constitution, problem, unilateral, Cofferati, concept, mankind, humanitarian, illegitimate, peace.

Thematic group 2 — References to the controversy about pacifism and its motivations (25.55% of the concordances).

in the streets, against, go down, demonstration, people, pace, liberation, demonstrate, Saddam, Iraq, believe, go, country, Roma, flag, pacifists, protest, person, cortege, world, motivation, pacifist, objective, April, to use, people, union, safe, Iraqi, right, be_for(pro)_, young people, government, to intend, test, opinion, to write, to continue, to free, campaign, dictator, to want, holy, national, example, to take, to support, to debate, struggle.

Thematic group 3 — References to American international policy and to the reasons for America' military intervention (32.52% of the concordances).

worldwide, second_world_war, lose, cold _war, crime, military, American, allies, to win, fourth, united_ states , to last, first_world_ war, man, enemies, to come, blood, year, power, effect, to hope, terror, terrorist, third, administration, to break out, history, month, century, Bush, terrible, to spend, menace, Americans, oil, politicians, enemy, fear, criminal, to respond, to know, front, Syria.

Thematic group 4 — References to war news and the Saddam regime (12.05% of the concordances).

civilians, dead, of_mass, gulf_war, victim, destruction, Afghanistan, weapons, during _the_war, child, town, Pentagon, soldier, bomb, bombing, hundred, to come back, Iraqis, missiles, soldiers, Iran, news, gulf, million, thousand, ten, westerners, damages, to strike, Baghdad, to arrive, propaganda, population, day, woman, journalist, prisoner, to die, death, troop, Vietnam, life, chief, TV, business, embargo, to tell, dirty, total.

response by the Bush administration, set inside the context of the oil issue and of the world wars of the 20th century.

The second major group is n.1, where we find the most frequent references to Italian foreign policy and to the main figures of national politics (Berlusconi, D'Alema, Fassino, Cofferati). Here, the debate deals with the questionable legitimacy of the American military intervention and with the critical position of the European countries.

Conclusion

As this report's results show, a content analysis of newsgroups generates several methodological and technical problems that are mostly due to the huge amount of textual data that need to be processed and to the presence of a remarkable amount of noise. The traditional techniques for reading texts are, in this case, quite useless or utterly wasteful. The first step is to use methods of quantitative automatic analysis that subsequently permit closer qualitative examinations.

This strategy of analysis turned out to be adequate enough to identify the main themes of the related debate in the messages of newsgroups in the site <it.politica> during the Iraqi war. In these messages we did not find distinct ideological and political trends in accordance with the thematic orientation of the newsgroups. An analysis of concordances did not reveal a specific lexicon about the war linked to the users' alleged ideological beliefs which, on the contrary, were connected more with the main themes that seem to be common across the same newsgroups. The users enjoyed engaging in a careful, informed, lively political debate characterized by strong argumentative accents. Yet, highly offensive verbal behavior occurred only occasionally and only in response to irrelevant provocations in the general environment. This result, given the strongly quarrelsome atmosphere caused by the event of the war, confirms the importance of the conversations occurring on web discussion groups for the development of political opinions.

An automatic content analysis that identified key words has highlighted three main themes in the discussions:

- The issues of the legitimacy of military intervention and the Bush administration's policies were debated; positive views went with motivations such as "liberation war," "holy war," "fair war," and "self defence war"; and negative ones were based on concepts related to "international illegality," and "war of aggression," or the danger of a nuclear war and an expansion of the war, up to forecasting a third or fourth world war.
- The debate over Italian Government political choices compared with other European countries; the most frequent references were to a preventive war and the Pope's statements, and the related religious and humanitarian implications.
- The debate on the controversy about pacifism and street demonstrations against the war.
- The debate on the war and its tragedy, the many dead among civilians, the bombs, the soldiers, TV news and war correspondents: this was a shared issue and it did not seem to be politically or ideologically oriented.

The results show that this methodology can be very effective in analyzing web interactions, and particularly at scanning political opinions and analyzing processes in the formation of public opinion. We only want to recall at this point that the Porto Alegre, Seattle, Okinawa, and Genoa movements, that later became the anti-globalization movement at an international level, grew strong as a political phenomenon thanks to the great opportunity provided by Computer Mediated Communication, by newsgroups, by forums and, generally speaking, by the possibility of communicating very quickly worldwide and without any traditional mediation.

The difficulty of analyzing thousands or even millions of messages generated in a period of one (or more) year/s has already been identified by some researchers (Fiore *et al.* 2002; Schobert *et al.* 2003). Such a difficulty, which is characterized by high background noise, can be overcome for the automatic mining of essential information. We can deal with this huge number of documents, which can be itemized according to key-words identifying their contents, by the automatic analysis of textual data and text-mining. Thus, we can select significant elements for in-depth and qualitative analyses.

While, judging by the international literature on the subject, little been done on the language of politics, the hints so far have been very promising. An analysis of a few newsgroups during the Yugoslavia secession (Stubbs 1998) has highlighted the role of CMC in political and cultural interaction. It also showed the strong ties among local, "real," and "virtual" communities. In another case study examining two very well-known newsgroups devoted to feminist culture (Fredrick 1999), the results revealed an environment of very close communication, characterized by a very sarcastic and accusatory style aimed at discriminating against dissenting opinions rather than including dissent. Completely different, but on the same wave-length as our report, are the results of a wide-ranging study (Pritchard 1999) that highlighted the extremely varied offering of themes and topics provided by Usenet, where users can find several virtual meeting places, topically focused ones, to discuss and confront issues in they are interested.

References

Bolasco, S. 1999. *Analisi Multidimensionale dei Dati.* Roma: Carocci.

Bolasco, S., F. Baiocchi and A. Morrone. 2000–2003. *TALTAC 1.6. Trattamento Automatico Lessico Testuale del Contenuto*, Roma: CISU.

Baym, N. K. 2002. "Il Posto delle Comunità Online Nella Vita Offline." *Rassegna Italiana di Sociologia* 43(1):55–71.

Day, R. 2001. *The Modern Invention of Information. Discourse, History, and Power.* Carbondale: Southern Illinois University Press.

Castells, M. 2000. *The Information Age: Economy, Society and Culture.* Oxford: Blackwell Publishers Ltd.

Choi, J. H. and J. Danowski. 2003. "Making a Global Community on the Net — Global Village or Global Metropolis? A Network Analysis of Usenet Newsgroups." *Journal of Computer Mediated Communication* 7(3). http://jcmc.indiana.edu/vol7/issue3/choi.html.

De Kerckhove, D. 1991. *Brainframes: Technology, Mind and Business.* Baarn: Bosch & Keuning.

De Kerckhove, D. 1995. *The Skin of Culture.* Toronto: Sommerville House Books.

Fiore, A. T., S. L. Tiernan and M. A. Smith. 2002. "Observed Behavior and Perceived Value of Authors in Usenet Newsgroups: Bridging the Gap." *Proceedings of the SIGCHI Conference on Human Factors in Computing Systems*, Minneapolis, USA.

Fredrick, C. A. N. 1999. "Feminist Rethoric in Cyberspace: The Ethos of Feminist Unsenet Newsgroups." *The Information Society* 15(3):187–197.

Galimberti, C. and C. Riva. 1997. *La Comunicazione Virtuale. Dal Computer alle reti Telematiche: Nuove Forme di Interazione Sociale.* Milano: Guerini e Associati.

Giuliano, L. 1999. "Comunità nel Ciberspazio. Comunicazione Elettronica e Spazio Sociale." *Sociologia. Rivista Quadrimestrale di Scienze Storiche e Sociali* 33(2): 47–63.

Giuliano, L. 2002a. "Identità e Narrazione in Ambienti Sociali Digitali." *Rassegna Italiana di Sociologia* 43(1):7–32.

Giuliano, L. 2002b. "G8-2001: la Rivolta nel Monitor. Analisi Testuale dei Messaggi nel Newsgroup <it.eventi.g8.genova> Durante gli Scontri di Piazza." *6° JADT*, 13–15 mars 2002, Saint-Malo, IRISA.

Giuliano, L. 2003. "L'analisi Automatica dei Testi ad Alta Componente di Rumore. Giornata di Studio su *Applicazioni di analisi testuale.*" Roma, 16 dicembre, Facoltà di Scienze Statistiche, Università Degli Studi di Roma, "La Sapienza".

Giuliano, L. 2004. "Il Lessico della Guerra nei Newsgroup della Categoria It. Politica Durante la Guerra in Iraq." 7° *JADT*, 10–12 mars 2004, Louvain-la-Neuve, UCL — Presses Universitaires de Louvain.

JADT. 2002. *6° Journées Internationales d'Analyse Statistiques des Données Textuelles*, IRISA, 13–15 mars 2002, at Saint-Malo.

JADT. 2004. 7° *Journées Internationales d'Analyse Statistiques des Données Textuelles*, Louvain-la-Neuve, 10–12 mars 2004, UCL — Presses Universitaires de Louvain.

Lebart, L., A. Morineau, M. Becue and L. Hausler. 1993. *SPAD-T 1.5 (DOS)*, CISIA.

Lévy, P. 1994. *L'intelligence Collective. Pour une Anthropologie du Cyberspace.* Paris: Éditions La Décourverte.

Lyman, P., K. Swearingen and H. R. Varian. 2003. *How Much Information?* The School of Information Management and Systems, Berkeley: University of California. http://www.sims.berkeley.edu/research/projects/how-much-info/internet.html>.

Nonnecke, B. and J. Preece. 2000. "Lurker Demographics: Counting the Silent." *Proceedings of the SIGCHI Conference on Human Factors in Computing Systems*, The Hague, Amsterdam.

Paccagnella, L. 1997. "Verso una Sociologia del Ciberspazio. Uno Studio di caso Sulla Conferenza Elettronica Cyber-punk." *Quaderni di Sociologia* 13:33–57.

Paccagnella, L. 2000. *La Comunicazione al Computer*. Bologna: Il Mulino.

Pritchard, G. S. 1999. "Placeless Publics: Identity, Community and Activity in the Usenet's Media." American Sociological Association conference paper.

Resnick, D. 1997. "Politics on the Internet: The Normalization of Cyberspace." *New Political Science* 41–42:47–67.

Rheingold, H. 1993. *The Virtual Community*. Reading: Addison-Wesley.

Rosen, D., J. Woelfel, D. Krikorian and G. A. Barnett. 2003. "Procedures for Analyses of Online Communities." *Journal of Computer Mediated Communication* 8(4). http://jcmc.indiana.edu/vol8/issue4/rosen.html.

Schobert, T., J. Preece and A. Heinzl. 2003. "Online Communities: A Longitudinal Analysis of Communication Activities." *Proceedings of the Hawaii's International Conference on System Sciences*, January 6–9, Big Island, Hawaii.

Smith, M. A. 1999. "Invisible Crowds in Cyberspace: Mapping the Social Structure of the Usenet." In *Communities in Cyberspace*, edited by M. A. Smith and P. Kollock. London: Routledge.

Smith, M. A. 2002. "Mapping Social Cyberspaces: Measures and Maps of Usenet, a Computer Mediated Space." *Dissertation Abstracts International, A: The Humanities and Social Sciences* 63(1):July, 382-A.

Smith, M. A. 2002. "Tools for Navigating Large Social Cyberspace." *Communications of the ACM* 45(4):51–55.

Stubbs, P. 1998. "Conflict and Co-Operation in the Virtual Community: EMail Ed the Wars of the Yugoslav Succession." *Sociological Research Online* 3(3). http://www.socresonline.org.uk/3/3/7.html.

Tuzi, A. 2003. *L'analisi del Contenuto*. Roma: Carocci.

Whittaker, S., L. Terveen, W. Hill and L. Chemy. 1998. "The Dynamics of Mass Interaction." In *Proceedings of Conference on Computer Supported Cooperative Work*. http://www.acm.org/pubs/citations/proceedings/cscw/289444/p257-whittaker/.

6 *Practices in the Use of ICTs, Political Attitudes Among Youth, and the Italian Media System**

Leopoldina Fortunati and Raimondo Strassoldo

Rationale

The interest of the survey presented here and the most important results obtained reside in the original research question to which we attempted to respond. What is the influence of the mainstream media (Meyrowitz 2004) on the opinion of young people in the Italy of the year 2002? It should be noted that there is an extraordinary concentration of media power in the hands of Silvio Berlusconi, who as President of the Council of Ministers has virtual control over RAI (the public TV channels) and, as an entrepreneur, possesses, through his family, a media empire made up of Mediaset (the second national TV pole), large publishing houses like Mondadori, dailies, weeklies, publicity agencies and so forth.

That the media have a certain amount of power to condition and orient public opinion (cf., for example, Noelle-Neumann 1984, 1991; Kourvetaris 1997; Bagdikian 2000; Neuwirth and Frederick 2004) is just common sense, even if this is felt intuitively much more than it is scientifically demonstrated. In fact, both the quantity and the modes of expression of this power, and above all its effects, are difficult to measure. The classic question "Can the media create and modify people's opinions, attitudes, and behavior?" is destined even in the most recent literature to remain unanswered for two reasons: the wide range

*The article has been discussed and revised by the two authors. Leopoldina Fortunati wrote pages 125 to 143, Raimondo Strassoldo wrote pages 144 to 152.

of variables involved (Braga 1992[1]; McQuail 1992) and the methodological limits that still afflict us today.

Furthermore, to make this ground even more difficult to explore is the fact that many forms of media (for instance, TV) are still relatively young, so the effects of their long-term consumption are impossible to gauge (Clark 1994). But apart from this, there is also another limit to consider, which is that longitudinal research, which is the only kind that can properly highlight the changes that have occurred during a determined period of time, are quite rare owing to the high costs involved. A similar situation, but even more restrictive, could be described for the computer.

However it must immediately be said that the difficulties that exist in measuring these effects, and therefore the partiality of the studies conducted so far, should not lead us to the conclusion that these effects do not exist or that they are not particularly relevant (Herman and Chomsky 1988; Popper 1994; Condry 1994; Derrida and Stiegler 1996; Bourdieu 1997; Couldry and Curran 2003). This conclusion, not difficult to find in the literature, leads to a serious logical flaw. It would be more correct to say, as Losito wrote in his book *Il potere dei media* (1994) (The Power of the Media), that the influence of mass communications on the public is actually mediated by such a high number of individual and social factors that it is impossible to check them all (always admitting that it is possible to describe them all).

In order to increase our understanding of this phenomenon, there have appeared over the years a series of analyses and theories, more or less properly conducted, that have attempted to explain this or that aspect of the problem (Chomsky 1994; De Fleur and Ball-Rokeach 1995). But none of them have been able to give a suitable answer to the question asked a few lines above. The point is that it is not that these theories are particularly weak, but that the question implies an answer that is so difficult and complex that it still remains unanswered. The influence of the media is mediated by the kind of use that is made of them, by the purpose for which they are used, by the context in which they are used, and by the gender, generation, class, educational and cultural typology of the audience, their styles of life, the personality of the users and so on (Silverstone 1994; Moores 1993). But even if we take this framework

[1]Braga (1992) stressed the variables connected to the "importance of the source," "contents of the message," and "audience predisposition." He also gave a survey of the variables connected with the overall mechanisms of the communicative process, such as "selective exposition, perception and memorization" and lastly the "intermediate factors" that can support or contrast with the formation of opinions.

for granted, the Italian situation is so peculiar because of the unusual concentration of media power in the hands of one person (the Premier) that we wondered — and this is our real research question — if by chance this particularly paroxysmal situation of media power was able to add anything new and significant to the same old question of the influence of the media and their role in democracy (McQuail 1992; Thompson 1995).

Aim, Sample, and Methods

Our first step was to reconstruct the use of the media as a background. Our focus was, on the one hand, to investigate the use and quantity of use of ICTs, such as TVs, computers, and other classic cultural instruments, such as books and magazines, among young people; and, on the other, the diffusion of modes of communicative sociality, also verifying if and in what terms practices in the use of ICTs correlate with practices in communicative sociality.

Our second step was to examine what views these young people have on certain socio-political issues on the agenda discussed by the mass media over these years, such as on one's own country (Anderson 1983; Pertierra 2004), peace, war, terrorism (Baudrillard 2002), and their judgment of the events of Genoa, to try to understand the most important dimensions of their attitudes on these crucial issues. The third step was to explore if and how young people's attitudes towards all of these issues correlate with their practices in use of ICTs and their modes of communicative sociality. This step is particularly important to answer the research question set forth in the rationale, that is, if the existence

[2]Here are more figures relating to the sample: 45.8% were made up of young people between the ages of 18 and 21, and 54.2% between the ages of 22 and 25; 34.7% are students, 14% are working students, 10% are unemployed, 30.4% are full-time workers and 5.3% temporary workers. As for education, 19.6% had little education (only primary), 9.4% a medium level (lower secondary), and 63.6% a high level (high school and university) (7.4% gave no answer). With regard to geographical area, 23.5% lived in the northwest, 16.5% in the northeast, 17.7% in the center, 28.9% in the south, and 13.3% on the islands (Sicily and Sardinia). As to the size of their hometowns, 33.3% lived in towns of up to 10,000 inhabitants, 22.4% in towns of from 10 to 30,000 inhabitants, 21.7% in towns of from 30 to 100,000 inhabitants and 22.6% in towns of over 100,000 inhabitants. As Clemente Lanzetti pointed out in his paper on the methodological aspects of this research (2002), "a status index has been constructed to analyze the distribution of subjects on a social scale and to see if there are significant differences in the replies of the interviewees due to their different positions on the scale itself. The index in this case served to synthesize various items of information, and give each person a score. The variables used for this operation were six and regard the professional qualifications and the level of schooling of the father, the mother and the interviewee. In drawing up the list of professions Max Weber's three components for status were referred to: power, prestige and economic aspect." The status index was divided in the categories of low, medium, and high.

of a "media regime" controlled by Premier Berlusconi, the so-called "Italian anomaly," has a strong conditioning effect on the public opinion of young people.

Our analysis is based on a representative sample of Italian youth (1,500 respondents aged 18 to 25 years, comprised of 764 males and 736 females; margin of error: 2.5%), interviewed by means of a questionnaire administered face to face in June 2002 (Strassoldo 2005).[2] The data were analyzed by means of descriptive techniques (frequency analysis) and the construction of contingency tables created by crossing observed variables with socio-demographic variables such as gender, age, education, activity, status, geographical area, size of hometown, and others such as political orientation and the dimension of religion. An inferential analysis was conducted by means of an χ^2 test. When the statistics of the χ^2 were significant, the analysis was developed by means of an examination of bi-varied log-linear models (Corbetta 1992). The purpose was to discover the associations between variable modalities that are at the root of the significance of the relation shown by the χ^2.[3] In some cases, t-test and models of one-way variance analysis were used. A factor analysis was applied to the opinions expressed by the 1,500 respondents on the main political issues that were taken into account. The exception was for the clashes in Genoa, which took place in July 2002 when thousands of people demonstrated against the policies expressed by the leaders of the countries belonging to G8. In this case, the study of the structure of relations between the responses obtained with socio-demographic variables and others such as political orientation and the religious dimension was again based on an analysis of bi-varied log-linear models.

The strong point of this research was the accuracy with which the sample was constructed, which made it possible to generalize the data to the Italian youth population aged between 18 and 25. The weak points were that of electronic media, since only TV and the computer were analyzed; and of printed media, where only books and popular magazines were examined. Other important media such as the radio, dailies, and so on, were excluded.

---

[3]In presenting the results, we will limit ourselves to referring to the significant points of the interaction between the two variables considered one at a time, and the standard values (points z) relative to the single cells to which a significant association parameter corresponds. In the text we have given in a note the z values with the indication of the significance level. An asterisk indicates a level of $p < 0.05$, two asterisks a level of $p < 0.01$, and three asterisks a level of $p < 0.0001$. We shall only give the figures that refer to the tables of intersections for which the Chi quadro test has given a significant relation.

However, this study is part of an ongoing line of research (Livolsi 1992; Fortunati 1998).

Results

Practices in the Use of TV Among Italian Young People

Almost the whole sample watches TV every day. They are subdivided roughly in the following way: one third watch TV for one hour (low consumption), one third from one to two hours (medium consumption), and the remaining third watch more than two hours (high consumption). The amount of time spent by young people in watching TV during workdays and holidays does not differ statistically (Table 1). Young people's TV consumption is therefore conditioned by various factors.

On the whole, while on holidays there is no gender difference, on workdays women watch more TV than men.[4] But the only specific difference that emerges is that during workdays more women than men of this age (67% vs. 60.6%) watch TV for more than one hour.[5] TV consumption during workdays is also influenced by activity, in the sense that students rank first among the young in having a TV consumption that is both medium (34.1%) and high (37.9%). More than half of the unemployed also reported having a high consumption of TV. On the other hand, swelling the number of non-viewers

Table 1. Hours of TV Viewing

Hours	Workdays	Holidays
Up to 1 hour	499 (33.3%)	456 (30.4%)
From 1 to 2	450 (30.0%)	404 (26.9%)
More than 2	500 (33.3%)	538 (35.9%)
No TV	49 (3.3%)	99 (6.6%)
No answer	2 (0.1%)	3 (0.2%)
Total	1500	1500

[4] $M = 4.07$ versus 3.80; $t_{(1496)} = -3.59$, $p < 0.001$.

[5] The analysis of log-linear models has shown the following significant relations between TV use and gender $\chi^2_{(2)} = 8.61$, $p < 0.05$, $z = 2.91^{**}$, activity $\chi^2_{(12)} = 62.51$, $p < 0.0001$, $z = 3.55^{**}$; $z = 2.17^{*}$; $z = 2.04^{*}$ e $z = 3.59^{***}$, TV use during workdays and geographical area $\chi^2_{(12)} = 31.23$, $p < 0.0002$, $z = 2.42^{**}$ e $z = 2.56^{**}$, on holidays $\chi^2_{(12)} = 29.09$, $p < 0.004$, $z = 2.88^{**}$ e $z = 2.12^{**}$, TV use during workdays and status $\chi^2_{(6)} = 16.21$, $p < 0.05$, $z = 2.82^{**}$, and TV use on holidays and status: $\chi^2_{(6)} = 14.00$, $p < 0.05$, $z = 2.06^{*}$ e $z = 2.52^{**}$.

is in proportion the number of temporary workers. It would seem that students and the unemployed have more time to dedicate to TV, so that a high consumption of TV seems to indicate a workday that is relatively stable or organized.

On TV consumption during workdays, geographical area also has quite an expected (it must be expected because this data confirm many other research data) influence. Among the respondents who indicated that their TV consumption is low, the highest percentage is concentrated in the Northwest (28.3% vs. 18.4% in the Northeast, 15.4% in the Center, 26.9% the South, and 11% the Islands), which is an area notable for the presence of large industries. Among the inhabitants of the islands (Sicily and Sardinia) the most widespread practice (40.7% of the local respondents) is quite a high level of TV consumption. The islands are also confirmed as an area of high TV consumption on holidays. In this they are joined by the south. Both are areas that are economically and industrially underdeveloped. Lastly, the amount of time spent in front of the TV screen depends on social status. From this research, not surprisingly, it emerges that for those with a low social status the most widespread practice both on workdays (40.8%) and holidays (41.1%) is to spend two hours or more in front of the TV every day. The unexpected fact is that on holidays, among those whose TV consumption is average, more than half are youth of high status. This reveals a picture of Sunday home life that is quite homely and sedentary even for young people from a background of strong cultural, social, and economic possibilities. In general, however, TV is a particularly lively resource in areas where a low income is associated with situations of social exclusion or geographical and/or cultural isolation. Neither age, education, nor size of hometown seem to condition TV use.

Obviously, what is important is not only the amount of the TV consumption, but also the contents to which young people are exposed. This is because the media go to create a doubling of reality, as shown by Luhmann (1996) and Baudrillard (1999). With regard to TV contents, the results illustrated in Table 2 show that young people essentially watch films, TV news (cf. Calabrese and Volli 1995), and sports. It is interesting to observe that more than half of young people do not follow information on TV at all (in part because, as we shall see later, they look for information on the internet; Magrini 2002). As for TV programs, young people's preferences are influenced by several factors, as we shall see below.

Starting from gender, we find that there is no gender difference as to TV news, documentaries, talk shows (Minnini and Ghiglione 1995), cartoons, and televideo. On the contrary, in general, men of this age watch much more

Table 2. TV Programs Chosen by Young People

Programs	First choice	2nd choice	No choice
Films	617 (41.1%)	344 (22.9%)	516 (34.4%)
TV News	329 (21.9%)	254 (16.9%)	894 (59.6%)
Sports	222 (14.8%)	202 (13.5%)	1053 (70.2%)
Fiction	76 (5.1%)	127 (8.5%)	1274 (84.9%)
Documentaries	76 (5.1%)	125 (8.3%)	1276 (85.1%)
Talk shows	49 (3.3%)	77 (5.1%)	1351 (90.1%)
Cartoons	39 (2.6%)	75 (5.0%)	1363 (90.9%)
Variety shows	34 (2.3%)	129 (8.6%)	1314 (87.6%)
Teletext	23 (1.5%)	44 (2.9%)	1410 (94.0%)
Quiz	19 (1.3%)	87 (5.8%)	1371 (91.4%)

This question was not answered by 23 respondents

sports than women, both as a first and second choice (26.9% vs. 2.9% and 22.1% vs. 5.1%). Women are more sensitive to narrative: in fact they watch, more than men, films as first choice (50.3% vs. 33.5%) and fiction as a first and second choice (8.5% vs. 1.9% and 14.5% vs. 2.8%). Again, there are more women among the audience that follows variety shows as a second choice (12.1% vs. 5.5%) and in general quizzes (9.2% vs. 5.2%).[6] This means that in TV consumption the difference in gender is not the result of a different interest on the part of men and women with regard to the informative function of TV, but relates to different preferences for the various entertainment programs. By articulating the analysis of TV programs with respect to other structural variables, it emerges that with regard to films (the type of program

[6]The analysis of log-linear models has shown the following significant relations: between gender and watching sports $\chi^2_{(2)} = 336.06$, $p < 0.001$, $z = 7.22^{***}$, $z = 3.01^{**}$ and $z = 15.31^{***}$; films $\chi^2_{(2)} = 46.78$, $p < 0.001$, $z = 4.80^{***}$ and $z = 6.21^{***}$; TV films $\chi^2_{(2)} = 111.30$, $p < 0.0001$, $z = 9.02^{***}$, $z = 2.47^{**}$ and $z = 3.54^{***}$; variety $\chi^2_{(2)} = 21.95$, $p < 0.001$, $z = 3.24^{***}$ and $z = 2.49^{**}$; quizzes $\chi^2_{(2)} = 9.61$, $p < 0.01$, $z = 2.76^{**}$; between TV news and age $\chi^2_{(2)} = 15.44$, $p < 0.001$, $z = 2.77^{**}$ and $z = 3.49^{***}$; political leaning $\chi^2_{(6)} = 27.56$, $p < 0.001$, $z = 2.24^{*}$; $z = 2.06^{*}$ and $z = 2.46^{**}$, between talk shows and education $\chi^2_{(4)} = 14.68$, $p < 0.01$, $z = 2.47^{**}$; $z = 2.36^{**}$, political leaning $\chi^2_{(6)} = 17.60$, $p < 0.01$, $z = 3.25^{***}$, between gender/age and film $\chi^2_{(6)} = 50.70$, $p < 0.0001$, $z = 4.86^{***}$; $z = 2.39^{**}$ $z = 2.11^{*}$ and 3.66^{***}, TV films $\chi^2_{(6)} = 115.77$, $p < 0.0001$, $z = 4.42^{***}$; $z = 4.02^{***}$ and $z = 2.36^{**}$, $z = 2.62^{**}$ and $z = 2.26^{*}$, variety shows $\chi^2_{(6)} = 23.10$, $p < 0.0009$, $z = -2.80^{**}$, sport $\chi^2_{(2)} = 6.94$, $p < 0.05$, between education and sports $\chi^2_{(4)} = 12.74$, $p < 0.05$, $z = 2.03^{*}$; $z = 2.53^{**}$, cartoons $\chi^2_{(4)} = 13.03$, $p < 0.05$, $z = 2.28^{*}$, between information use of TV and gender/age $\chi^2_{(3)} = 13.80$, $p < 0.01$, $z = 2.57^{**}$ and $z = 2.32^{*}$, political leaning $\chi^2_{(3)} = 11.04$, $p < 0.05$, $z = 3.05^{**}$, and computer use $\chi^2_{(2)} = 16.55$, $p < 0.001$, $z = 2.95^{**}$ and $z = 3.82^{***}$.

overwhelmingly the most followed by young people), the girls of both age groups give them as their favourite programs more than do the boys (29% and 30.5% vs. 19.8% and 20.7%). The situation is similar for TV films (Liebes and Katz 1988), even if they are followed much less by our respondents (only 13.6% give them as their first or second choice). Women make up 82.5% of their audience: in particular, girls of both age groups are an overwhelming majority among the few respondents (76 in all), who give it as first choice, while above all the more grown-up girls (42.5%) give it as second choice.

As for news (which however, it must be remembered, is seen only by 38.8% of the respondents), it is the young people from 23 to 26 that give it in most cases (26.1%) as a first choice, as opposed to the younger ones from 19 to 22 (17.7%) of whom 64.7% do not watch it at all, despite the increase in the narrative quality of the news (Bird and Dardenne 1988). Still on the subject of news, those of different political leanings also have a different interest in the news.[7] In fact, 68.8% of young people who are centrists and 66.8% who are apolitical (and they are the highest percentages) never follow TV news. Among those who give it as a second choice there are proportionally more young people of the right (20.7%). With regard to TV debates (watched only by 8.4% of young people), one variable that is important is education: the vast majority of those who have a low level of education express a total disinterest in talk shows, while among the few that give them as their favourite programs more than half have a medium level of education. To complete the picture, we add that almost all of the young people who are uninterested in politics avoid it completely.

By putting together preferences for TV news, teletext (consulted only by 4.4% of respondents), documentaries (followed by 13.4%), and debates (8.4%), with the aim of determining the overall use of TV for obtaining news, the picture so far described is reconfirmed. The use of TV to obtain information is much higher among the more grown-up young people, both males (30.6%) and females (30.4%), than with the younger ones from 18 to 21 years of age (respectively 20.6% and 18.3%). This means that TV information has little meaning below a certain age (22 years). In fact it is more functional to the world of work, family and social responsibilities, and civil society, while the

[7]These are the frequency figures for political leaning and the importance of the religious dimension among respondents. It emerged that 27.1% are left-wing and 27.7% are right-wing; only 7.4% are centrists, while the highest percentage (37.2%) is make up of those who are not interested in politics. As for the religious dimension, 7.5% consider it fundamental to their lives, 24.2% important, 33.4% quite important, 23.9% not very important, and 10.4% irrelevant.

world in which adolescents and young people live tends to concentrate rather in a more familial and local sphere. They are still at that stage in which they have to learn and reinforce their secondary socialization in environments that are strongly rooted in the surrounding territory, such as school, the district, etc. As for politics, an analysis of the use of TV for information reinforces what was said before, that three quarters of those who do not like politics never use TV for that purpose. By examining, on the other hand, how the use of the various media is mixed and combined by young people, we observe that among those who use TV for information the highest percentage (38.5%) is constituted of young people who usually make great use of the computer, while 71.3% of those who make little use of it, but whose use is aimed at obtaining news and information, do not use the TV at all for this purpose. So there is a consistent percentage of young people who just have a quick look at information on the internet, while another segment uses both means of communication intensively in a synergistic manner, probably to extract the best from both modalities.

Variety shows, on the other hand, clearly have little attraction for young people (they are watched by only about 10.9% of the respondents), but are rejected even more by girls between the ages of 18 and 21, most likely because of the stereotyped image of the women depicted in such shows. With respect to sports (seen in general, remember, by less than one third of the respondents) it emerges that three quarters of the youngest respondents never watch sports programs on TV (probably because they are more involved in playing the sports directly), while one third of the older young people cite them as their first or second choice of program. It is interesting to observe that those with a medium level of education make up about one-fifth (but this is the highest percentage) of those who state that sports are their favorite program, while three quarters of the more educated respondents do not give sports programs as their first choice. Lastly, it is again the education variable that distinguishes the behavior of the respondents with respect to cartoons: the less educated are in proportion the most numerous (8.3%) among the few that give them as their second most favorite type of program. Therefore, TV sports and cartoons are reconfirmed as the programs that manage to entertain young people of culturally lower levels.

Practices of Computer Use in Italy

In Italy 71.6% of young people between the ages of 18 and 25 use a computer. For the first time, there is no significant gender difference in computer use (cf. Fortunati 1998, but also at a European level, Mante-Meijer *et al.* 2001). This means that among young people 18–25 years of age the gender divide for this instrument has disappeared. Neither is there any difference in the use of this technology between the younger (19–22) and older (23–26) respondents. Many other factors, on the other hand, continue to differentiate the use of this instrument. For example, different degrees of education mean significant differences in practices of use: young people with a high level of education use the computer more than others (80.4% vs. 46.3% and 58.9%).[8] Status is another variable that can differentiate computer use: over half of those with a low level of education do not use it, while 87.7% of those with a high level do. In computer use, the size of one's hometown also has a significant influence: it is in towns with over 100,000 inhabitants that the percentage of young people who use the computer is higher (78.7%), while it is lower in towns of from 30,000 to 100,000 inhabitants. With regard to specific parts of the country (Nielsen distinguished the following areas: Northeast, Northwest, Centre, South, Islands), it emerges that the area where the computer is used significantly less by young people between the ages of 18 and 25 is the South (63.4%), and where it is used most is Central Italy (77.4%). In the end, political leanings also affect computer use, in the sense that more than three quarters of left- and right-wing youth use the computer, while more than a third of young people not interested in politics do not. To conclude, the factors that seem to influence the persistence of the digital divide are, apart from social, cultural, and political and economic ones, also those connected with the different development of urban infrastructure.

There are several reasons why the computer is used by young people in Italy between the ages of 18 and 25: play, study, search for news, work, and as pastime (Table 3). What is noticeable is that, while the number of young people who use the computer for play or to search for news decreases in relation to the length of time dedicated to these activities, the study has, on the contrary, a positive progression. But of all of the reasons, let us isolate for reasons of space, the data concerning the use of the computer to search for news. As can be seen from Table 3, almost three quarters of the young people who use the computer use it also or only for the purpose of obtaining information.

[8]The analysis of log-linear models has shown the following significant relations: between computer use and education $\chi^2_{(3)} = 135.18$, $p < 0.0001$, $z = 7.68^{***}$, $z = 2.62^{**}$, $z = 6.80^{***}$; status $\chi^2_{(3)} = 137.42$, $p < 0.0001$, $z = 11.03^{***}$, $z = 8.88^{***}$; size of hometown $\chi^2_{(3)} = 12.40$, $p < 0.01$, $z = 2.11^{*}$ and $z = 3.28^{***}$, and geographical area $\chi^2_{(4)} = 25.12$, $p < 0.0001$, $z = 1.98^{*}$ and $z = 4.30^{***}$.

Table 3. Computer Use

Duration of use	Play	Study	News	Work	Pastime
Never	429 (40.7%)	330 (31.2%)	302 (28.7%)	499 (48.2%)	175 (16.5%)
Less than one hour	336 (31.8%)	196 (18.6%)	429 (40.7%)	115 (11.1%)	338 (31.9%)
One hour	182 (17.2%	196 (18.6%)	214 (20.3%)	94 (9.1%)	284 (26.8%)
More than one hour	108 (10.2%)	334 (31.6%)	108 (10.2%)	326 (31.5%)	263 (24.6%)
Total	1055	1056	1053	1034	1060

The number of people using the computer drops in inverse proportion to the time dedicated to looking for news on the internet. So there is light use of the computer as a source of information. The intersection with the structural variables tells us that the map on this specific use is quite articulated: almost half of the young women (19–22) — and it is the highest percentage — are concentrated in medium use category (one hour a day).[9] The unemployed and 40.7% of the young people from the South instead, in greater proportion than other groups, declare heavy use (from one hour upwards), even if it must be pointed out that more than half of the young people who make great use of the computer are of medium status. As for the specific use of the computer for the purpose of obtaining information, it is not used in this way by one third of the respondents from hometowns of under 10,000 inhabitants. To this must be added more than one third of the young people with little education or low status. Our reading of the data suggests that the computer can constitute an important source of information for young people in search of employment or who are trying to shake off geographical and/or cultural isolation, as long as they have at least some strong resources from their family background.

By crossing the use of the computer with the TV, where in both cases the purpose is to seek news, it turns out that more than one third of the young people who make regular use of the computer for information have the habit of also using the TV for that purpose, while almost half of those who do not use the TV for information are light users of the computer as a means of obtaining information. So apart from the hard core of young people who

[9]The analysis of log-linear models has shown the following significant relations: between computer use for news and gender/age $\chi^2_{(6)} = 14.48$, $p < 0.05$, $z = 2.15^*$, education $\chi^2_{(6)} = 16.78$, $p < 0.05$, $z = 2.19^*$ activity $\chi^2_{(8)} = 16.06$, $p < 0.05$, $z = 2.25^*$, size of hometown $\chi^2_{(6)} = 14.10$, $p < 0.05$, $z = 2.36^{**}$, status $\chi^2_{(4)} = 11.29$, $p < 0.05$, $z = 2.73^{**}$, $z = 2.20^*$, geographical area $\chi^2_{(8)} = 25.01$, $p < 0.01$, $z = 4.01^{***}$, political leaning $\chi^2_{(3)} = 32.61$, $p < 0.0001$, $z = 2.56^{**}$, $z = 2.37^{**}$, $z = 4.50^{***}$, and TV use for news $\chi^2_{(2)} = 16.55$, $p < 0.0001$, $z = 3.82^{***}$, $z = 2.95^{**}$.

want to be informed and also check information in various ways, many more young people have shifted to the internet, which they are using as a source of obtaining information rapidly.

Practices of Reading Books and Magazines and Listening to Music

Listening to music is the most widespread behavior among young people, as is reported in Table 4. Three quarters of respondents do so often, while a significant percentage indicated that they read books and magazines sometimes (44.2% for books and 49.5% for magazines). Music seems to have found a new energy in the new technologies and has expanded in an extremely transversal way among the young, showing once again that it is the most universal language that exists.

By crossing these three variables with socio-demographic variables, media behavior variables, and attitudes, an interesting map emerges of the media-consumption behavior of young people. By starting with the listening to music, it turns out that 81.7% of those who live in urban centers dedicate themselves to this activity often, while half of those who never listen to music (24 respondents in all) live in hometowns that do not have more than 10,000 inhabitants.[10] As to status, the great majority of those of high status listen to music regularly, while the highest percentage of those who never listen to music (6%) are of low status. Political attitudes can also condition the listening of music: the majority of those who are left-wing declared that they listen to music often, while nearly half of those who listen to it only sometimes stated that they are not interested in politics. To conclude, listening to

Table 4. Book and Magazine Reading and Music Listening among Young People

Programs	Books	Magazines	Music
Never	293 (19.5%)	499 (33.3%)	47 (3.1%)
Sometimes	663 (44.2%)	743 (49.5%)	311 (20.7%)
Often	537 (35.8%)	246 (16.4%)	1135 (75.7%)
No answer	7 (0.5%)	12 (0.8%)	7 (0.5%)
Total	1493	1488	1493

[10]The analysis of log-linear models has shown the following significant relations: between listening to music and hometown $\chi^2_{(6)} = 13.10$, $p < 0.05$, $z = 2.39^{**}$; $z = 2.43^{**}$, status $\chi^2_{(4)} = 14.73$, $p < 0.01$, $z = 2.90^{**}$; $z = 2.42^{**}$, and political leaning $\chi^2_{(6)} = 17.54$, $p < 0.01$, $z = 2.03^{*}$; $z = 2.79^{**}$.

music is an extraordinarily widespread activity among young people and the few who do not listen to it show signs of their social isolation in the fact that they, more than others, are from small hometowns, have low status, and are quite indifferent to politics.

But how is the activity of reading resisting the advance of the electronic media? In Italy, which is a country that is less keen on reading than other industrially advanced European countries, how much do the young people read? Half of the young people in Italy are still students and so, as a matter of course, can be expected to be more involved in reading (ISTAT 1998; Fortunati and Manganelli 1998). From an examination of reading, it emerges that among the non-readers, three quarters are men, while 64.2% of keen readers are women.[11] In particular, more than one third of the younger male respondents and almost a third of the less young ones never read books. On the contrary, nearly half the women from the age of 23 to 26 years read a book once every now and then, while the other half, accompanied by 42.4% of the younger ones, dedicate themselves to reading often. This research therefore confirms a fact that is already well-known, which is that reading is an activity engaged in by women more than by men. But if we continue with the analysis of the other structural variables, we discover that education, activity, status, and size of hometown also weigh on the propensity to read. Almost half of those who never touch a book are not very well educated, while the majority of those who read with regularity have a high level of education. It is, however, interesting to observe that in Italy more than one third of those with a medium level of education never read books, showing that the phenomenon of a return to illiteracy might also affect those levels of the population that have a medium level of education. Among the non-readers we find in the first place full-time workers (46.2%) (followed at a distance by temporary workers and the unemployed), while it is students, also those who work, who in almost half of the cases are frequent readers of books. Also a small hometown is associated

[11]The analysis of log-linear models has shown the following significant relations: between reading books and gender $\chi^2_{(2)} = 128.32$, $p < 0.0001$, $z = 9.46$***; $z = 10.04$***, gender/age $\chi^2_{(6)} = 133.84$, $p < 0.0001$, $z = 6.73$***; $z = 6.02$***, $z = 2.19$*, $z = 4.09$*** and $z = 7.38$***, education $\chi^2_{(6)} = 191.62$, $p < 0.0001$, $z = 8.10$***; $z = 4.24$***, $z = 7.54$***, activity $\chi^2_{(8)} = 145.38$, $p < 0.0001$, $z = 4.92$***; $z = 2.60$**, $z = 4.34$*** and $z = 7.48$***, size of hometown $\chi^2_{(6)} = 31.53$, $p < 0.0001$, $z = 3.36$***; $z = 2.50$**, $z = 2.33$** and $z = 3.56$***, status $\chi^2_{(4)} = 137.02$, $p < 0.0001$, $z = 9.71$***; $z = 9.28$***, TV use for news $\chi^2_{(2)} = 21.50$, $p < 0.001$, $z = 3.08$**; $z = 4.61$***, computer use for news $\chi^2_{(4)} = 37.59$, $p < 0.0001$, $z = 5.84$***; $z = 2.20$*, $z = 2.71$** and $z = 2.23$*, and political leaning $\chi^2_{(6)} = 50.98$, $p < 0.0001$, $z = 4.92$***; $z = 3.15$***, $z = 3.90$***.

with a low propensity to read: 41% and more than a quarter of non-readers are from hometowns with up to 10,000 or 30,000 inhabitants, respectively. By contrast, of those from larger hometowns (over 100,000 inhabitants) half read books occasionally and the other half regularly. Non-readers are associated with low status in 40.8% of cases and, vice versa, more than half of the young people with high status are keen readers. These data confirm that factors such as the scarcity of cultural instruments, belonging to a low status bracket, the heaviness of a full workday or the lack of it, and living in a small hometown, are associated — as was easy to predict — with a low propensity to read.

Crossing with ICT use reveals an articulated relationship between TV and computer use for information purposes and reading activity. Let us start by saying that among non-readers three quarters are uninterested in following the news on TV and half do not use the computer to obtain information. But at the same time, nearly half of those who watch TV and make heavy use of the computer to obtain news are keen readers. More specifically, among light users of TV to obtain information almost one half read sporadically, the other half frequently. In conclusion, the habit of reading is also connected to a certain interest in following the news both on TV and the computer, which is why, with regard to the news, the use of these three technologies is reinforced rather than excluded. Lastly, there is also a significant connection between political leaning and reading: nearly half of the leftist young people read often, while more than half of the centrist respondents read only occasionally. Furthermore, among non-readers, almost half are not interested in politics at all.

If this is the reading behavior among the respondents of our sample, what is the situation with activities that involve less engaged reading, that is, that involve the products of mass culture such as popular magazines? Among the socio-demographical variables, only three — education, activity, and status — can differentiate reading behavior with regard to popular magazines. Let us start with the first: more than one third of the young people with little education and with low status never read them,[12] while more than half of the

[12]The analysis of log-linear models has shown the following significant relations: between reading popular and educational magazines $\chi^2_{(6)} = 15.59$, $p < 0.05$, $z = 3.16^{***}$, activity $\chi^2_{(8)} = 15.66$, $p < 0.05$, $z = 2.58^{**}$; $z = 2.51^{**}$, status $\chi^2_{(4)} = 15.80$, $p < 0.01$, $z = 1.96^{*}$; $z = 3.32^{***}$, TV use for news $\chi^2_{(4)} = 11.72$, $p < 0.05$, $z = 2.38^{**}$; $z = 2.42^{**}$, TV use during holidays: $\chi^2_{(6)} = 27.56$, $p < 0.001$, $z = 2.52^{**}$; $z = 2.92^{**}$ $z = 1.99^{*}$, computer use for information purposes: $\chi^2_{(4)} = 18.21$, $p < 0.01$, $z = 3.41^{***}$; $z = 2.40^{**}$ and $z = 2.47^{**}$, and computer use for pastime: $\chi^2_{(4)} = 9.85$, $p < 0.05$, $z = 2.05^{*}$; $z = 2.69^{**}$.

students, 44.2% of the temporary workers and 58% of those with high status confess to reading them occasionally. The scarcity of cultural and economic resources, therefore, as was easily predictable, is associated with less of a propensity to read not only books, but also popular literature. However, there was no automatic association between reading this kind of publication and readers with a low level of education and low status.

As for the use of ICTs, more than half of average users of TV sometimes happen to read popular magazines, but as many as 39.4% and 41.5% of frequent readers are made up respectively of heavy users of TV and average users of TV specifically during holidays. At the same time, almost half of those who never watch TV and a little more than half of the light users of TV, again during holidays, never read them. The use of computers also differentiates this popular magazine-reading behavior, in the sense that 39.5% of those who do not use the computer to search for news do not read them, while 57.2% of light computer users and one-fifth of heavy computer users read them with this aim respectively sometimes and often. Instead, more than half of the heavy users of the computer as a pastime often read them, while 40% of the non-users of the computer as a news source do not read them. We may conclude by saying that we note a clear convergence between the reading of popular magazines and the regular use of TV and the computer probably because they are used in some cases to relax from more important activities, and in others as a different form of entertainment that is synergic with the other forms.

Practices of Communicative Sociality

The most widespread practices of communicative sociality among young people in Italy are reported in Table 5.

The most common activities are being with friends, going to the disco, the bar, clubbing and, to a lesser extent, sports (Eurobarometer 2004).[13] The activities engaged in less assiduously are shopping, excursions, and going to the cinema and to shows. Those activities that are far less widespread are political activities and trade union activities, parish activities, volunteer service, hobbies, the theater and, in general, creative artistic activities. But how is the picture of the practices of sociality among Italian youth structured?

[13]We define sports as a social practice of communication because among the principal benefits of sports expressed by respondents in the European survey "The Citizens of the European Union and Sport" is the fact of being with friends (31%). Moreover, in 2004 sports continued to interest more men and young men than women in Europe (Eurobarometer 2004), exactly as we found in this Italian survey.

Table 5. Practices of Communicative Sociality

Activity	Sports, gym	Excursions	Being with friends	Theater	Shopping	Artistic and creative activities	Hobbies	Volunteer service
Base	1500	1500	1500	1500	1500	1500	1500	1500
Never	287 (19.1%)	714 (47.6%)	14 (0.9%)	946 (63.1%)	202 (13.5%)	888 (59.2%)	1060 (70.7%)	1072 (71.5%)
Sometimes	633 (42.2%)	655 (43.7%)	243 (16.2%)	476 (31.7%)	869 (57.9%)	388 (25.9%)	324 (21.6%)	322 (21.5%)
Often	578 (38.5%)	115 (7.7%)	1235 (82.3%)	65 (4.3%)	417 (27.8%)	212 (14.1%)	102 (6.8%)	95 (6.3%)
Missing	2 (0.1%)	16 (1.1%)	8 (0.5%)	13 (0.9%)	12 (0.8%)	12 (0.8%)	14 (0.9%)	11 (0.7%)

Activity	Bar, clubbing, etc.	Political, trade union, cultural activities	Parish activities	Disco	Cinema, shows	Others
Base	1500	1500	1500	1500	1500	1500
Never	139 (9.3%)	1258 (83.9%)	1108 (73.9%)	136 (9.1%)	175 (11.7%)	90 (6.0%)
Sometimes	599 (39.9%)	178 (11.9%)	283 (18.9%)	550 (36.7%)	909 (60.6%)	10 (0.7%)
Often	752 (50.1%)	50 (3.3%)	93 (6.2%)	804 (53.6%)	404 (26.9%)	39 (2.6%)
Missing	10 (0.7%)	14 (0.9%)	16 (1.1%)	10 (0.7%)	12 (0.8%)	1361 (99.3%)

How are the social communicative practices associated and harmonized by young people, who at the same time make quite lively use of these means of communication and information? Let us say immediately that for reasons of space our analysis will obviously concentrate on the most widespread forms of communicative sociality among young people, in the hope of finding some interesting indications to explain the relations that are being established today in the social body between mediated and non-mediated aspects of communicative practices among Italian youth.

Let us start by analyzing activities related to spending time with friends, a practice — let us remember — that is covered in 83.5% of cases with "regularity." More young men than women (86.4% vs. 79%) spend time with friends often,[14] confirming in this a practice that is most common among men of a more socialized style of life. With regard to age, the only significant association is related to those who spend time with friends only sometimes: among them, 65.8% belong to the 23–25 year group. So, as time passes and they become more adult, young people lose the capacity for steady relations with friends. With regard to status and activity, it emerges that among those who never spend time with friends, half are of a low status and more than half are full-time workers. This means that a scarcity of resources and regular work hinders the possibility of being with friends. Regarding the use of ICTs, heavy users of the computer for study often find more time to spend with friends than others, while users of TV for purposes of information have less time to spend with friends. Thus, a "serious" use of the computer is compatible with being with friends, while the use of TV for purposes of information may lead to a certain de-socialization.

Another widespread practice of communicative sociality is going to the disco. More women than men (58.9% vs. 49.2%) declare that they go rarely or never, reinforcing yet again the picture of a less socialized female life. As to age, it emerges that it is young people from the age of 23 to 26 more than others (58.1% vs. 49.6%) who never or rarely go to the disco; in particular, one third of those who never go to the disco are made up of women from the age of 22 to 25.[15] Therefore, the more you become an adult, the less you go to the disco. Also the size of one's hometown has an influence on this social communicative practice. It is in fact definitely less common in small hometowns (under 10,000 inhabitants), while nearly one third of respondents who go to the disco very frequently (that is, more than once a week) live in medium-to-large towns (that is, over 100,000 inhabitants). Thus, going to the disco turns out to be more of an urban than rural custom, and looking

[14]The analysis of log-linear models has shown the following significant relations: between spending time with friends and gender $\chi^2_{(2)} = 14.18$, $p < 0.0001$, $z = 2.08^*$, age $\chi^2_{(2)} = 15.85$, $p < 0.0001$, $z = 2.09^*$, status $\chi^2_{(4)} = 10.97$, $p < 0.05$, $z = 2.12^*$, activity $\chi^2_{(8)} = 21.53$, $p < 0.01$, $z = 2.02^*$, computer use for study purposes $\chi^2_{(4)} = 10.26$, $p < 0.05$, $z = 2.05^*$ and TV use for news $\chi^2_{(2)} = 8.23$, $p < 0.05$, $z = 2.55^{**}$.

[15]The analysis of log-linear models has shown the following significant relations: between going to the disco and gender $\chi^2_{(2)} = 14.81$, $p < 0.001$, $z = 3.68^{***}$, age $\chi^2_{(2)} = 10.08$, $p < 0.01$, $z = 3.01^{**}$, gender/age $\chi^2_{(6)} = 26.25$, $p < 0.001$, $z = 4.11^{***}$, size of hometown $\chi^2_{(6)} = 16.80$, $p < 0.05$, $z = 2.50^{**}$ and $z = 2.23^*$, geographical area $\chi^2_{(8)} = 16.70$, $p < 0.05$, $z = 2.58^{**}$, TV use $\chi^2_{(4)} = 10.32$, $p < 0.05$, $z = 2.16^*$ and $z = 2.40^{**}$, TV use during workdays $\chi^2_{(6)} = 15.34$, $p < 0.05$, $z = 2.90^{**}$, $z = 2.12^*$ and $z = 2.63^{**}$, political leaning $\chi^2_{(6)} = 20.65$, $p < 0.01$, $z = 2.24^*$.

at geographical area, it is in proportion more common among young people who live in central Italy than in other parts of the country (13.3% vs. 10.6% in the Northwest, 8.5% in the Northeast and 7.2% in Sicily and Sardinia). As for TV consumption, the results show that the more you go to the disco the less you watch TV: in particular, nearly half of light TV users (less than one hour a day) go to the disco more than once a week, while more than half of heavy TV users (more than two hours a day) go to the disco rarely or never. This trend is also confirmed by examining workdays only. In the end, as in politics, the only significant relation is with left-wing youth, who in more than half of the cases (and it is the highest percentage) never (or rarely) go to the disco.

Yet another practice, spread over half of the sample with "regularity," is going to the bar or clubbing. Among assiduous bar-frequenters are significantly more men than women (56% vs. 45.1%), reconfirming once again the less socialized life style of women. In particular, in first place are found young males between 18 and 23 (57.8%) followed by older ones (53.9%).[16] Nearly one third of sporadic frequenters and non-frequenters of bars and clubs live in small and medium-sized (from 30,000 to 100,000 inhabitants) towns, while more than half of the inhabitants of medium/large towns go to them habitually. Geographical area is also a variable with a great influence on this practice of communicative sociality: more than half of assiduous frequenters live on the Islands, while 40.3% of Southerners and nearly a quarter of those who live in central Italy never go. Of the young inhabitants of the North-west, 41.3% do go sometimes. Therefore, going to the bar or a club is an urban more than a rural habit, and is especially common in highly industrialized areas and the Islands (perhaps to overcome the geographic isolation). With regard to ICT use, it emerges that more than half of assiduous frequenters on the one hand do not watch news on TV and, on the other, make heavy use of the computer as a pastime.

As for sports (distributed among young people in 50.1% of cases often and 38.5% sometimes): in general, more young men than women participate in sports often (45.3% vs. 31.6%), while among non-sporting types women take

[16]The analysis of log-linear models has shown the following significant relations: between going to bars/clubbing and gender $\chi^2_{(2)} = 16.87$, $p < 0.001$, $z = 3.74^{***}$, gender/age $\chi^2_{(6)} = 26.06$, $p < 0.001$, $z = 1.98^{*}$ and $z = 2.22^{*}$, size of hometown $\chi^2_{(6)} = 18.64$, $p < 0.01$, $z = 2.07^{*}$, $z = 3.59^{***}$ and $z = 2.39^{**}$, geographical area $\chi^2_{(8)} = 33.29$, $p < 0.001$, $z = 2.50^{**}$, $z = 2.72^{**}$, $z = 3.10^{***}$, $z = 2.30^{*}$ and $z = 2.20^{*}$, TV use for news $\chi^2_{(2)} = 9.63$, $p < 0.001$, $z = 3.12^{***}$ and $z = 2.46^{**}$, computer use as pastime $\chi^2_{(4)} = 13.28$, $p < 0.05$, $z = 3.09^{***}$, and political leaning $\chi^2_{(6)} = 16.38$, $p < 0.05$, $z = 3.07^{**}$ and $z = 3.41^{***}$.

first place. In particular, this greater propensity on the part of males to engage in sports involves young men of both of the age groups considered, while a refusal to engage in sports activities is higher among the younger women (almost a quarter of younger women never take part in sports, which is the highest percentage).[17] Probably a lower propensity on the part of women for sport is reinforced by a lifestyle that is, as we have seen, less socialized. Swelling the ranks of the non-sports lovers more than others are temporary workers (28.7%) and full-time workers (24.1%). On the contrary, almost half of students engage in sports sometimes, and another 41.3% participate in sports often, preceded in this by worker-students (44%). Increasing the number of non-sporting people by one-third are those with low status, while 44.9% and 45.9% of those with high status, more than the others, engage in sporting activities occasionally and frequently, respectively. The highest concentration (more than one-fifth) of non-sportsmen/women is located in small/medium towns, while 43.4% and 45.7% of those who live in medium/large towns, more than the others, engage in sports sometimes and often, respectively. Evidently, the urban environment leads to greater involvement in sports activities, perhaps because it is an environment that imposes less daily movement. With regard to the use of the computer, almost half (47.3%) of those who never take part in sports are young people who never use the computer for study purposes, while more than half of those who engage in sports sometimes are heavy users of it for study purposes. Among the non-sporting types, the greatest number (40.7%) is made up of young people who never use the computer to search for information. Thus, computer use is associated with a certain disinclination to engage in sporting activities only when such use is heavy. As regards political leanings, right-wing youth engage in sports more often than others (45.3% vs. 36.5% of those on the left, 33.9% in the center, and 36.2% who are apolitical), while among the non-sporting young people the highest percentage are apolitical (45.4%).

[17]The analysis of log-linear models has shown the following significant relations: between engaging in sports and gender $\chi^2_{(2)} = 31.43$, $p < 0.001$, $z = 5.56^{***}$; $z = 3.26^{***}$, gender/age $\chi^2_{(6)} = 35.84$, $p < 0.0001$, $z = 4.67^{***}$; $z = 1.98^{*}$, $z = 2.30^{*}$, activity $\chi^2_{(6)} = 34.31$, $p < 0.0001$, $z = 2.17^{*}$; $z = 3.13^{***}$, $z = 2.71^{**}$, $z = 2.61^{**}$ e 2.24^{*}, status $\chi^2_{(4)} = 50.72$, $p < 0.0001$, $z = 6.15^{***}$; $z = 3.25^{***}$, $z = 5.13^{***}$, size of hometown $\chi^2_{(6)}$=29.88, $p < 0.0001$, $z = 2.31^{*}$; $z = 2.19^{*}$, $z = 2.02^{*}$, $z = 3.81^{***}$, computer use for study purposes $\chi^2_{(4)} = 24.27$, $p < 0.001$, $z = 4.55^{***}$; $z = 2.16^{*}$, $z = 2.24^{*}$, computer use for information: $\chi^2_{(4)} = 13.55$, $p < 0.01$, $z = 3.68^{***}$, and political leaning $\chi^2_{(6)} = 18.22$, $p < 0.01$, $z = 3.29^{***}$; $z = 2.41^{**}$.

Computer and TV Use and Opinions about One's Country, War, Peace, and Terrorism

After delineating the practices in the use of the various means of information and communication, and the social communicative practices and the relations that exist between the former and the latter, we went on to assess the fundamental dimensions of young people's attitudes to political issues so greatly debated in the world of information, such as one's country, peace, war, and terrorism. We also tried to determine whether these dimensions of attitude are in some way influenced by structure variables, forms of communicative sociality and, above all, the use and contents of the media.

To obtain these opinions, the respondents were asked the following questions: (1) Recently, the Italian government — starting from President Ciampi — has been speaking increasingly of patriotism, respect for the national flag, singing the national anthem, etc. How far do you agree with the following statements? (2) Recently there has been much talk of the legitimacy or not of war, and how to build the peace. How far do you agree with the following statements? (3) What are the conditions for the possibility in the future of abolishing wars and arriving at a real and lasting peace? (4) Last July, there were violent clashes between demonstrators and the police. What is your opinion of these events? (5) In your opinion, were the United States and their Allies right to carry out military actions against terrorism after September 11? Each question was proposed together with a battery of items (one's country: 8 items; peace: 7 items; war: 10 items; terrorism: 9 items). For each item, the respondents had to express their degree of agreement/disagreement on a scale of 4 points, where 1 = I do not at all agree, and 4 = I am in total agreement.

The data were elaborated by applying to the matrices of correlation among the items factorial analysis with the main components method.[18]

One's Country

Let us begin from young people's attitude towards their country. Two factors emerge from a factorial analysis: the first is defined as "refusal of

[18]The factors that emerged from the various factorial analyses were then rotated by the Varimax system and Kaiser's normalization process. Therefore, for each dimension the composite factorial scores were calculated by considering the items with saturations higher than 0.400 in one single factor. The reliability of each of the factors was then assessed by means of Crombach's α coefficient. The relation between these psychological variables was then studied with behavioral variables (consumption of means of communication and forms of communicative sociality), with other attitudinal variables (referring to politics and religion), and with some structural variables, by applying Student's t-test for gender and univariate variance analysis with a between-factor for all of the other variables.

patriotism;"[19] the second, which has more of an emotional character, expresses an "appreciation of national symbols."[20] As for the whole sample, the overall factorial score for the first factor is 1.88 (d.s.= 0.56), and for the second 2.52 (d.s. = 0.71). This means that, with regard to the first dimension, the respondents share the refusal of patriotism to a small degree, but at the same time, in relation to the second factor they show that they have a certain appreciation of national symbols.

In relation to these two factors (which together explain 52.53% of the total variance), we must say immediately that there are no differences in the scores for the amount of TV seen daily. Apart from this, as regards the first factor F1, the profile of those who share the refusal of patriotism is mostly composed of those who have a low level of education[21] (in fact, refusal of patriotism is strengthened as the level of education drops), those of medium status, those who live in the Northeast and Northwest, those who live in towns of more than 100,000 inhabitants, those who never engage in sports, those who never use a computer, those who go to the theater often, those who never do shopping, those who do not take part in the life of the parish, those who consider the religious dimension irrelevant, those who give debates as

[19]The first factor F1 saturates the items: these manifestations are only a kind of consolation for the loss of real sovereignty and national identity, owing to the European Union, globalization, etc. = 0.733; these manifestations are only tactics to conceal the real problems of Italy, such as unemployment, injustice, exploitation, etc. = 0.732; these manifestations are a counter-balancing of federalist policies, regional autonomy, etc. = 0.732; these manifestations of nationalism in Italy are only an indication of the return of the old values of the right = 0.602; and this is only a passing fashion = 0.512.

[20]The second factor F2 saturates the items: at last in Italy, too, we can be proud of our own country = 0.804; for too long patriotism, the national flag, the national anthem, etc. were neglected by the old parties = 0.753; when I see the national flag fluttering, hear the national anthem being sung and played and I think of my country, I am overwhelmed by emotion = 0.736. For F1 the alpha coefficient is 0.68 and for F2 is 0.70.

[21]With regard to the first factor F1, the results of a one-way variance analysis with between the factor of education are as follows: education $F_{(3,1489)} = 4.34$, $p < 0.01$; status $F_{(2,1490)} = 3.46$, $p < 0.05$; geographical area $F_{(4,1488)} = 3.65$, $p < 0.01$; size of hometown $F_{(3,1489)} = 3.95$, $p < 0.01$; sports $F_{(2,1488)} = 5.01$, $p < 0.01$; computer use $F_{(1,1489)} = 9.46$, $p < 0.01$; go to theater $F_{(2,1477)} = 3.05$, $p < 0.05$; go shopping $F_{(2,1488)} = 6.84$, $p < 0.01$; participate in parish life $F = 14.21$, df 2,1474, $p < 0.001$; religious dimension $F_{(4,1481)} = 8.79$, $p < 0.001$; talk shows $F_{(2,1469)} = 3.12$, $p < 0.05$; and political leaning $F_{(3,1479)} = 19.14$, $p < 0.001$.

As to the second factor F2, geographical area $F_{(4,1488)} = 14.62$, $p < 0.001$; size of hometown $F_{(3,1489)} = 4.31$, $p < 0.01$; sports $F_{(2,1488)} = 6.84$, $p < 0.01$; popular magazines $F_{(2,1479)} = 8.89$, $p < 0.001$; shopping $F_{(2,1478)} = 10.11$, $p < 0.001$; creative activities $F_{(2,1478)} = 5.89$, $p < 0.01$; parish life $F_{(2,1474)} = 11.80$, $p < 0.001$; religious dimension $F_{(4,1480)} = 16.59$, $p < 0.001$; talk shows $F_{(2,1467)} = 3.32$, $p < 0.05$; TV use for information $F_{(1,1468)} = 5.09$, $p < 0.05$; cartoons $F_{(2,1467)} = 6.09$, $p < 0.01$; films $F_{(2,1467)} = 3.91$, $p < 0.05$; and political leaning $F_{(3,1479)} = 14.12$, $p < 0.001$.

their first choice of TV program, and those who are left-wing or do not have a precise political orientation. So the first profile of an urbanized youth, who is from the North, has an average income, is sedentary, lay, politically engaged, informed, and culturally lively, even if with a low educational profile.

As for the second factor F2 (a positive attitude towards national symbols), the profile of those who take this attitude is of a person who lives in the South, in a small/medium-sized town, often engages in sports, often reads popular magazines, often does the shopping, never does creative work, participates in parish activities and considers the religious dimension fundamental, gives TV debates as his/her second choice, but who on the whole also uses TV to obtain information and who belongs to the Center-right. This second profile of young people is in some ways opposite to the first, in that it is of a type from the South, rural, religious, informed, politically tending to the Right and more disengaged politically, but socially active.

Peace

Let us now consider the attitude towards peace (Giuliano 2004). Here, two factors emerge: the first factor can be defined in terms of "peace as the abolition of the causes of war,"[22] and the second sees "education towards peace" as the most effective strategy to follow in order to eliminate war.[23] These two factors together account for 49.8% of the total variance. As for the whole sample, the composite factorial score is 2.22 (d.s. = 0.47) for the first factor and 2.78 (d.s. = 0.38) for the second. This means that the respondents maintain, with regard to the first dimension, that they do not believe so much in the concrete possibility of removing the causes of war in order to promote peace, but instead are quite agreed in considering the educational strategy important.

As for the first factor F1, here too, TV consumption does not make any difference, while what does make a difference are other factors. More than others it is those who go to the cinema rarely or never as well as those who

[22]The first factor F1 saturates the following items: on the condition of the economic and social equality of all men in the whole world = 0.756; on the condition of the abolition of cultural and religious differences = 0.666; on the condition of the end of capitalist exploitation = 0.630; on the condition of forbidding the production of arms and the abolition of armed forces = 0.592; on the condition of the abolition of states, of the unification of humanity and one world government = 0.432.

[23]The second factor F2 saturates the following items: on the condition of education for peace, brotherhood, solidarity, and universal love = 0.856; and on the condition of the abolition of racial inequality = 0.730. For F1 the alpha coefficient is 0.61 and for F2 it is 0.52. Thus, we must bear in mind that this second factor has a slightly weak level of reliability.

give cartoons as their second choice among their favourite TV programs, who believe that to promote peace the right strategy is to remove the causes of war. Therefore, it would seem that those who believe more in the removal of the social and economic causes of war are young people with a propensity for iconic narration. With respect to the second factor, the longer the time that is spent daily in front of the TV,[24] the more it is thought that there is a possibility that the causes of war can be removed to promote peace. In the same way, those who use TV for the specific purpose of obtaining information, those who use the computer, who read books often, who have left-wing political leanings, are more convinced of the validity of the educational strategy. Therefore, the profile of those who favor more education on peace as an antidote to war is young people on the left, who are quite well-informed and of a good cultural level.

War

With regard to young people's attitude to war, two factors emerge: the first can be defined as "refusal of war,"[25] and the second as "war as an inevitable evil."[26] By verifying the reliability of each of the two factors by means of Crombach's α coefficient, it emerged that for F1, the alpha coefficient turned out to be 0.72, while for F2 it was 0.62. Overall, the two factors together

---

[24]With regard to the first factor F1, the results of the one-way variance analysis with between factor cinema are as follows: $F_{(2.103)} = 3.52$, $p < 0.04$; cartoons $F_{(1.107)} = 6.69$, $p < 0.05$. As to the second factor F2, the results are: TV use $F_{(2.105)} = 4.40$, $p < 0.05$; TV use for information $F_{(1.1468)} = 5.09$, $p < 0.05$ computer use $F_{(1.107)} = 5.45$, $p < 0.03$; reading books $F_{(2.104)} = 4.52$, $p < 0.02$; political leaning $F_{(3.104)} = 3.68$, $p < 0.05$.

[25]The first factor F1 saturates the following items: we must always be against every war = 0.764; just wars do not exist; all wars are unjust = 0.727; history shows that wars only cause death and destruction, and create more problems than they solve = 0.707; the threat of external enemies must be combated with diplomacy, dialogue, openness, being open to compromise = 0.654; wars happen above all because there is economic and social injustice.; and to abolish wars we must abolish all forms of inequality among men = 0.492.

[26]The second factor F2 saturates the following items: in the course of history wars have also had positive effects, for example by freeing oppressed peoples and destroying criminal political regimes, such as Nazism = 0.751; at times there are situations in which it is necessary to use armed forces for humanitarian interventions, in defence of civil rights, to protect against catastrophes, etc. = 0.714; when a nation is threatened or attacked by enemy forces from outside, it has the right to defend itself also with arms = 0.611; in the course of history, the needs of war and the armed forces have stimulated scientific, technical, and organizational progress = 0.559; wars happen only because there are armies and armed forces; and to abolish wars, and bring peace to the world, it would be enough to abolish armed forces = −0.442 (the scores of this item have been inverted).

explain 44.9% of the total variance. As for the entire sample, it turns out that the composite factorial score for the first factor F1 is 3.06 (d.s. = 0.58), while for the second factor F2, it is 3.00 (d.s. = 0.48). This means that some of the respondents maintain that it is quite right to refuse war, while others think fatalistically that war is a necessary evil.

Those who reject war are mostly women[27] (thus continuing the traditional feminine hostility towards war), those who live in small/medium-sized towns, who live on the Islands, who are keen readers, who listen to music frequently, who regularly take part in artistic activities who rarely or never go to the disco, who do not give sports as their favourite TV programs, who do not play with the computer, who use the computer for the purpose of study often, who often or sometimes do volunteer work, who often participate in the life of the parish, who consider the religious dimension important or fundamental to their lives, who work in politics quite frequently, and who have left-wing leanings. The profile of those who are against war is therefore constituted of women, young people from a rural background, from the Islands, who take themselves seriously, Catholics but of the Left, and those who are socially and culturally active.

Those who tend to maintain that war is a necessary evil are men,[28] in particular those between 22 and 25 years, who live in the Northwest, are politically oriented to the right, often meet up with their friends, often go to the bar or clubbing, go to the disco several times a week, often engage in sports, often go in for political activities, never read, never listen to music,

[27]With regard to the first factor F1, the results of the t-test are: $t_{(1477)} = -5.69$, $p < 0.001$, while the results of a one-way variance analysis with between factor geographical area are as follows: $F_{(4,1474)} = 2.61$, $p < 0.05$; size of hometown $F_{(3,1475)} = 2.78$, $p < 0.05$; books $F_{(2,1470)} = 5.26$, $p < 0.01$; music $F_{(4,1470)} = 4.31$, $p < 0.05$; artistic activities $F_{(2,1466)} = 6.98$, $p < 0.01$; disco $F_{(2,1467)} = 8.64$, $p < 0.01$; sport programs $F_{(2,1456)} = 4.67$, $p < 0.01$; computer for play $F_{(2,1040)} = 3.13$, $p < 0.05$; computer for study $F_{(2,1041)} = 3.54$, $p < 0.05$; volunteer work $F_{(2,1466)} = 6.54$, $p < 0.05$; parish life $F_{(2,1461)} = 3.46$, $p < 0.05$; religious dimension $F_{(4,1467)}F = 5.25$, $p < 0.001$; and political leaning $F_{(3,1467)} = 37.84$, $p < 0.001$.

[28]With regard to the first factor F2, the results of the t-test are: $t_{(1484)} = 7.78$, $p < 0.0001$, while the results of a one-way variance analysis with between factor gender/age are as follows: $F_{(3,1482)} = 20.72$, $p < 0.001$; geographical area $F_{(4,1481)} = 2.84$, $p < 0.05$; political leaning $F_{(3,1476)} = 29.71$, $p < 0.001$; sport $F_{(2,1481)} = 7.31$, $p < 0.01$; books $F_{(2,1477)} = 5.45$, $p < 0.01$; music $F_{(2,1476)} = 3.30$, $p < 0.05$; friends $F_{(2,1476)} = 4.03$, $p < 0.05$; theater $F_{(2,1471)} = 7.47$, $p < 0.01$; creative activities $F_{(2,1473)} = 4.78$, $p < 0.01$; volunteer work $F_{(2,1473)} = 5.92$, $p < 0.01$; bar, clubbing $F_{(2,1475)} = 4.27$, $p < 0.05$; political activities $F_{(2,1470)} = 3.24$, $p < 0.05$; disco $F_{(2,1475)} = 6.73$, $p < 0.05$; sports on TV $F_{(2,146)} = 4.68$, $p < 0.01$; variety $F_{(2,1462)} = 3.66$, $p < 0.05$; computer use for study $F_{(2,1048)} = 3.72$, $p < 0.05$.

never go to the theater, are not accustomed to dedicating themselves to creative activities, never do volunteer work, give sports programs on TV as their first and second choices, never watch variety programs on TV, and use the computer to study less than one hour a day or never. The profile that emerges is of an adult male, who is a right-wing militant, is sports-minded, is sociable, and has a low cultural level.

Strategies Against Terrorism

Examining the attitudes of young people regarding strategies to use against terrorism three factors emerge: the first factor is one that we will call "the Afghanistan war as the only solution to terrorism";[29] the second is what we might call "the real reasons (economic and for revenge) of the Afghanistan war";[30] and the third is "terrorism is to be fought with counter-espionage and the international police."[31] When we verified the reliability of each of the three factors by means of Crombach's α coefficient, for F1 the alpha coefficient turned out to be 0.74, for F2 is 0.68, and for F3 is 0.23. So the third factor, not having an acceptable level of reliability, will not be considered in the following analyses. The first and the second factors together explain 48.4% of the total variance. The composite factorial scores are, for the first factor, 2.38 (d.s. = 0.68) and for the second 2.25 (d.s. = 0.70). This means that a larger portion of the respondents quite agree with the view that the Afghanistan war was the only possible solution to terrorism, but also that another part of respondents quite agree with the view that the Afghanistan war was dictated by economic reasons, by revenge, and to sanction political and military superiority.

With respect to the first factor F1, it emerges that those who are in greater agreement than others with the view that the Afghanistan war was the best

[29]The first factor F1 saturates the items: the oppressive and criminal Taleban regime must be destroyed by every means = 0.826; there was no other way to destroy the terrorist groups = 0.824; terrorism is beaten by abolishing injustice = −0.619; and we should have used diplomacy and carried on a dialogue with the Taleban government = −0.589 (the scores of these last two items have been inverted).

[30]The second factor F2 saturates the following items: the Afghanistan war was waged only to show the superiority of the western/Christian world over the Muslim one = 0.831; the Afghanistan war only satisfies a desire for revenge = 0.798; the Afghanistan war was waged only for reasons of economic strategy (oil pipelines, etc.) = 0.637.

[31]The third factor F3 saturates the following items: terrorism is fought only with counter-espionage, the police, etc. = 0.732; in Afghanistan it was not a war, but only an international police operation = 0.724.

response against terrorism were men,[32] young people aged 22 to 25, those with a lower level of education, the unemployed and temporary workers, those who live above all in the Northwest, who live in towns of medium size (from 30,000 to 100,000 inhabitants), who are politically right-wing, who often meet up with their friends, who go to bars and clubbing, who go to the disco several times a week, who often engage in sports, who never do volunteer work, who never do militant work for a political party, who never or rarely go to the cinema, who never go to the theater, who never take part in any creative activities, who never read books, who listen to music sometimes, who do not follow talk shows on TV, who give as their first choice TV sports programs, who give films as their second choice, and who do not use the computer even for study or to search for news. The profile that is delineated by this factor is that of a young adult, with little education, with work problems, from the North, who is quite urbanized, leans politically to the right, is very sociable, is culturally and technologically backward, and poorly informed.

With respect to the second factor F2, it emerged that those who are most convinced that the Afghanistan war was dictated by precise economic reasons, reasons of revenge, as well as to sanction political and military superiority[33] were working students, those who live in the Northeast, those who live in medium/large-sized towns, those who often go on trips, who often go to the theater, who go to the cinema more than once a week, who take parts artistic activities more often than most, who read often, who make great use

[32] As to the first factor F1, the result of the t-test is: $t_{(1476)} = 8.93$, $p < 0.001$, while the results of the one-way variance analysis with between factor gender/age are as follows: $F_{(3,1474)} = 27.74$, $p < 0.001$; education $F_{(3,1474)} = 6.46$, $p < 0.001$; activity $F_{(4,1470)} = 7.26$, $p < 0.001$; geographical area $F_{(4,1473)} = 3.05$, $p < 0.05$; size of hometown $F_{(4,1473)} = 3.05$, $p < 0.05$; political leaning $F_{(3,1467)} = 59.60$, $p < 0.001$; friends $F_{(4,1468)} = 4.15$, $p < 0.05$; bar/clubbing $F_{(2,1467)} = 4.13$, $p < 0.05$; disco $F_{(2,1465)} = 7.69$, $p < 0.001$; sport $F_{(2,1473)} = 3.34$, $p < 0.05$; volunteer work $F_{(2,1466)} = 9.43$, $p < 0.001$; political activity $F_{(2,1462)} = 4.11$, $p < 0.05$; cinema $F_{(2,1463)} = 8.36$, $p < 0.001$; theater $F_{(2,1463)} = 13.25$, $p < 0.001$; creative activities $F_{(2,1464)} = 20.86$, $p < 0.001$; books $F_{(2,1469)} = 30.03$, $p < 0.001$; music $F_{(2,1469)} = 3.88$, $p < 0.05$; talk shows $F_{(2,1455)} = 4.35$, $p < 0.05$; sport on TV $F_{(2,1455)} = 12.96$, $p < 0.001$; films $F_{(2,1455)} = 4.43$, $p < 0.05$; computer use $F_{(1,1474)} = 9.20$, $p < 0.01$; computer use for work $F_{(2,1041)} = 9.83$, $p < 0.001$; and computer use for searching for news $F_{(2,1038)} = 4.12$, $p < 0.05$.

[33] With regard to the second factor F2, the results of the one-way variance analysis with between factor activity are as follows: $F_{(4,1461)} = 3.65$, $p < 0.01$; geographical area $F_{(4,1464)} = 2.65$, $p < 0.05$; size of hometown $F_{(3,1465)} = 5.24$, $p < 0.01$; excursions $F_{(2,1458)} = 5.78$ $p < 0.01$; theater $F_{(2,1453)} = 8.66$, $p < 0.001$; cinema $F_{(2,1454)} = 3.69$, $p < 0.03$; creative activities $F_{(2,1455)} = 4.95$, $p < 0.01$; books $F_{(2,1459)} = 4.30$, $p < 0.02$; computer use for study $F_{(2,1037)} = 4.95$, $p < 0.01$; shopping $F_{(2,1455)} = 7.22$, $p < 0.01$; volunteer work $F_{(2,1457)} = 3.02$, $p < 0.05$; parish life $F_{(2,1458)} = 3.33$, $p < 0.05$; political leaning $F_{(3,1458)} = 32.28$, $p < 0.001$; religious dimension $F_{(2,1457)} = 3.73$, $p < 0.01$; and political activity $F_{(2,1453)} = 11.51$, $p < 0.001$.

of the computer for study, who never go shopping, who do volunteer work more often than most, who do not take part in the life of the parish, and who consider the religious dimension not very important or irrelevant to their daily lives, who are left-wing and are active politically often or at least sometimes. Thus, the profile of those who believe that the war in Afghanistan was dictated by reasons completely different from those that were is that of a young person who is a working student, lives in the Northeast, is very urbanized, active, socially committed, secular, politically left, and very lively culturally.

The Events of Genoa

With regard to the clashes in Genoa, which took place in July 2002 when thousands of people demonstrated against the G8 leaders, the young people's opinions were ambivalent. Of the sample, 71.3% stated that police behavior was correct on that occasion and 66.9% that the police violence was justified. But, at the same time, 72.6% said that the police were not professional enough in handling the events that took place, 66.5% that this was the reason that they hit out at everyone indiscriminately, and 77.3% that in any event the police were wrong to strike young demonstrators. One thing, however, is certain: for 88.5% of the young people interviewed, the events of Genoa were not exaggerated by the media. What were the factors that conditioned these opinions? Let us proceed in order and start from the conviction expressed by the majority of interviewees (71.3%) that the police behaved correctly. Of the young people who hold firmly to this conviction, 86% are right-wing, 76.1% quite religious, 77.3% of low status; while among those (nearly a third) who think the opposite, almost half are left-wing, more than third not at all religious, and about one third of high status and urbanized. If we go on to analyze the conviction expressed by 66.9% of the sample that the police violence was unwarranted, it emerges that over half of the left-wing youth, almost half of the a-religious youth (44.2%), and the young people with university degrees (45.5%) and urbanized youth (39.6%), think exactly that the police violence was unwarranted. On the contrary, 86.9% of right-wing youth, 71.5% of those who live in small towns, and 69.2% of those with a medium level schooling maintain that the police violence was justified. Going on now to analyze the conviction shared by 72.6% of the sample that the police were unprofessional and not up to the task, it turns out that 78% of the left-wing youth took this position, 76.2% of girls, 77.9% of those of high status, 84.8% of those young people who live on the islands, and 76% of the urbanized youth; while those who claim that the police were up to the task included 35.4% of the right-wing youth, almost one-third of boys (29.7%), about one-third of those of low status (31.7%), those who live in central Italy (30.7%) and those who live in rural centers (30%) and about one-fifth (22%) of those who live in centers with from 30,000 to 100,000 inhabitants. Analyzing now the assertion shared by 77.3% of the sample that "the police hit out indiscriminately," it turns out that 78% of left-wing youth, 73.9% of those who live in towns, and 69.3% of those who do not use TV for information, share this claim; while almost half of the right-wing youth, 39.1% of urbanized young people in medium-to-small towns (from 10,000 to 30,000 inhabitants), and 37.3% of those who use TV for information think, on the contrary, that the police did hit out indiscriminately. This is the only case in which TV consumption for the purpose of obtaining information turns out to be associated with a conviction by the respondents that the case corresponds to the official version of events as given by the government. As for the consideration, shared by 77.3% of the young people interviewed, that the police had in any case been wrong to beat up the young people, it emerges that 38.7% of the right-wing youth, 27.1% of the young people of low status, and 29.5% of the young people with a low level of schooling (and they are always the highest percentages in relation to the other categories of response) are convinced of the contrary. Lastly, concerning the conviction, shared by 88.5% of the sample, that the Genoa events were not an exaggeration by the media, it emerges that 94.3% of the left wing youth and 95.5% of the a-religious share this belief. Furthermore, among those who think this, 64% are made up of the less young (22–25 years old). On the contrary, 91.5% of the younger ones (18–21 years) and almost one-fifth (17.9%) of the right-wing youth are convinced of the contrary. On the whole, we may conclude that the use of means of communication such TV or the computer definitely had little bearing on the opinions of young people in Italy on the events of Genoa; rather, political and religious convictions, status, degree of urbanization of home town, and level of schooling were the important factors. The weight of variables such as gender, age group and geographical area were less significant.

Conclusions

This research draws above all an interesting picture of, on one hand, the use of TV and the computer, as well as music, books, and popular magazines; and on the other hand, practices of communicative sociality. Television emerges on the whole as a particularly active resource in those areas where a low income is associated with situations involving the risk of social exclusion or geographical and/or cultural isolation, and possibly an organized workday relatively stable. We must not, however, forget that only one half of the world of young people is in some way affected by TV information. This limited exposure to TV information does not denote differences of gender, but shows an interesting generational split between those in the age group of 18 to 21 years, who turn out to be quite uninterested in news, and those between 22 and 25, who shows a certain interest towards the world of information. This split is not surprising, because information is employed in the world of work, family and social responsibilities, and civil society. Instead, the horizon of the world in which adolescents and young people live is necessarily largely familial and local. They are still in the stage of life in which they have to understand and reinforce their secondary socialization in environments that are strongly rooted in the surrounding territory, such as school, the neighborhood, etc. It is this specific condition of theirs that leads them, more than adults, to perceive the dimension of present public events as extraneous.

Even more surprises are revealed with regard to the computer. Perhaps the most relevant fact is that 2002 signals the transition from a period marked by a clear digital divide between men and women to one in which this discrepancy has been completely overcome, at least with the age group of from 18 to 25 years. Another aspect still is that the computer appears to be an instrument for information that is important for young people, who in half the cases are shifting to the internet and using it as a source of information for rapid consultation. The computer is especially important for young people in their search for employment or attempts to overcome geographical and/or cultural isolation, as long as they have some strong resources from their family environment.

Music seems to have found a new strength in the new technologies to expand transversally among the young, showing once again that it is the most universal language. Listening to music is a form of cultural behavior that is extraordinarily widespread among young people. The few who do not listen to it show signs of social isolation and, more than others, live in small hometowns, have a low status, and are quite indifferent to politics.

As for reading, this research reconfirms that factors such as the scarcity of cultural instruments, having low status, the heaviness of a full workday or a lack of it, and living in a small town, are associated — as was easy to predict — with a low propensity to read. Furthermore, it highlights an already known fact, which is that reading is an activity that involves more females than males. Lastly, this research underlines how the habit of reading is also related to a certain interest in following the news both on TV and the computer, so that these three technologies, if filtered through a transversal interest in news, are mutually strengthened, rather than exclusive. And the scarcity of cultural and economic resources turns out to be associated, as was easily predictable, also with a low propensity to read popular literature. However, there is no automatic association between the reading of this kind of publication and readers with little education and low status. In fact, there is no clear convergence between the reading of popular magazines and even the sustained use of TV and the computer, probably because they are used in some cases during moments of relaxation from more taxing activities, and in others as a different form of evasion that becomes synergic with the other forms.

An analysis of the practices of communicative sociality also confirms an already well-known fact, but we may be permitted to have doubts as to its persistence: the practice of a socialized life being more widespread among men. This trend does, however, persist: women spend less time with friends, go less to the disco, go less to the gym, go less to bars and clubs, and so on. Another well-known element of the dynamics of communicative sociality is the fact that the more time that passes and we become adults, the less often we will meet with friends, go to the disco, and so forth. Lastly, another quite well-known element is that the "serious" use of the computer may be compatible with seeing friends, while the use of the TV also for purposes of information risks developing into a situation of greater de-socialization.

If this is the picture of the use of ICTs and the practices of communicative sociality among young people, what is their effect on their political attitudes? In particular, has this examination of the use of TV and the computer in Italy, in the present context of an increasing concentration of media power, been able to highlight any new elements with respect to the perennial question of the influence of the media?

Let us briefly reconsider the profiles of the young people with regard to the various political dimensions that have emerged. Those who reject patriotism a little more than the others and identification with its symbols are urbanized young people, from the North, with an average income, who do not move much, a secular, politically committed, informed, and culturally lively, even if with a low educational profile. By contrast, those who have positive attitude towards national symbols are Southern young people, rural, religious, informed, politically right-wing and less committed culturally, but socially active. Television seems important therefore for supporting the building of a strong national identity more than for criticizing the drift towards nationalism.

Those who believe more than others that the right strategy to promote peace is to remove the social and economic causes of wars are young people with a propensity towards iconic narration. Those who tend to show greater support for education on peace as an antidote to wars are young persons with leftist leanings, who are well-informed and of a high cultural level. Therefore, TV is important also for supporting the idea of the importance of education for peace, while the computer is not.

The profile of those who are against war is made up of women, rural young people, Island dwellers, the very serious-minded, Catholics who are also leftist, and those who are socially and culturally committed. With respect to war, those who consider it a necessary evil are adult males, right-wing militants, and sportsmen, who are sociable and culturally not very accomplished. On the opinions common among young people concerning war, neither the TV nor the computer seems to be completely non-influential.

The young people who consider the Afghanistan war to be the only solution against terrorism are between 22 and 25 years of age, have little education, and have employment problems; they are Northerners, come from small hometowns, are politically oriented to the right, take part in a wide range of social activities and are popular, are culturally and technologically backward, and little informed. Those who instead interpret the Afghanistan war as dictated by reasons completely different from those given officially are young people who are working students, those who live in the Northeast, are from medium-sized hometowns, are active, socially engaged, secular, politically to the left, and very active culturally. It may be interesting to note that the two dimensions that emerge on strategies against terrorism — the Afghanistan war as the only solution to terrorism and the Afghanistan war as dictated by reasons quite different from those given officially — while representing opposing viewpoints, are characterized by virtually absent TV information (or deriving from the internet).

Lastly, as regards the events of Genoa, the profile of those who approve of police behavior on that occasion is made up mainly of those politically oriented to the right, who are religious, of low status and cultural attainment, who are poorly informed, and who are from rural areas. On the other hand, among the less numerous ranks of those who disapprove of the behavior of the police on that occasion are left-wing, a-religious, of high status and education, and urbanized. Again, the influence of TV information on these two different attitudes towards the events of Genoa is marginal.

From the analysis of all of the data we collected, we may sum up by underlining at least four elements: (1) the greater hostility of women to wars;

(2) the double soul of the Catholic world constituted of both young people of the left and right;[34] (3) the weight of the rural/urban infrastructure in the modeling of diffusion and use of electronic media (De Certeau *et al.* 1981); and (4) the fact that television (and the computer) have had only a marginal influence on the political attitudes of the young. The present results belie the views of those who had hypothesized that the so-called "Italian anomaly" would directly strengthen the political consensus on the part of the young on government programs being carried out. On the whole, the use of the media has little relation with what young people think politically. Thus, we find that not even Berlusconi's media empire is able to condition young public opinion. What creates public opinion is not so much the direct influence of the various media, but rather the overall informational environment, where the media are important actors, certainly, but not the only ones. As many writers on the subject have shown (Lazarsfeld *et al.* 1944; Katz and Lazarsfeld 1955; Katz 1957), to be effective, a single medium must be mediated by other factors, such as the family, school, the social environment, religion, politics, and other media (Meyrowitz 2003).

References

Anderson, B. 1983. *Imagined Communities: Reflexions on the Origin and Spread of Nationalism.* London: Verso.

Bagdikian, B. 2000. *The Media Monopoly.* Boston: Beacon Press.

Baudrillard, F. 2003. *L'esprit du Terrorisme.* Paris: Editions Galilée.

Bird, E. S. and R. W. Dardenne. 1988. "Myth, Chronicle, and Story: Exploring the Narrative Qualities of News." In *Media, Myths, and Narratives. Television and the Press,* edited by J. W. Carey. London: Sage.

Bourdieu, P. 1997. *Sulla Televisione.* Milano: Feltrinelli.

Braga, G. 1992. "Le Capacità Persuasive del Messaggio." In *Sociologia Delle Comunicazioni di Massa,* edited by G. Fabris. Milano: Angeli.

Calabrese, O. and U. Volli. 1995. *I telegiornali. Istruzioni per l'uso.* Roma-Bari: Laterza.

Chomsky, N. 1994. *Media Control. The Spectacular Achievements of Propaganda Manufacturing Consent: Media and Propaganda,* tr.it. Il Potere dei Media. Firenze: Vallecchi.

[34] Actually, the Catholic world corresponds to the large majority of the sample: 78.9%. Inside this world, one quarter is left-wing, one third is right-wing, and 36.8% are a-political. On the other hand, 93.7% of young people of the political center are Catholic, while atheists and agnostics, who are in general very few (11%), are more highly concentrated (46.7%) among the ranks of the left $\chi^2_{(6)} = 57.78$, $p < 0.0001$, $z = 3.95^{***}$; $z = 3.31^{***}$.

Clark, C. S. 1994. "La violenza in TV." In *Cattiva Maestra Televisione*, edited by K. R. Popper and J. Condry. Milano: Reset.

Corbetta, P. 1992. *Metodi per le Scienze Sociali*. Bologna: Il Mulino.

Couldry, N. and J. Curran, eds. 2003. *Contesting Media Power*. Lanham: Rowman and Littlefield.

Condry, J. 1994. "Ladra di Tempo, Serva Infedele." In *Cattiva Maestra Televisione*, edited by K. R. Popper and J. Condry, Milano: Reset.

De Certeau, M., P. H. Jeudy, and P. Virilio, eds. 1981. *Tecnologie del controllo. Il Quotidiano Urbano*. Roma: Theorema.

DeFleur, M. L. and S. J. Ball-Rokeach. 1995. *Teorie delle Comunicazioni di Massa*. Bologna: Il Mulino.

Derida, J. and B. Stiegler. 1996. *Échographies de la Télévision*. Paris: Éditions Galilée.

Eurobarometer. 2004. *The Citizens of the European Union and Sport*.

Fortunati, L., ed. 1998. *Telecomunicando in Europa*. Milano: Angeli.

Fortunati, L. and A. M. Manganelli. 1998. "La Comunicazione Tecnologica: Comportamenti, Opinioni ed Emozioni degli Europei." In *Telecomunicando in Europa*, edited by L. Fortunati Milano: Angeli.

Giuliano, L. 2004. "L'analisi del Contenuto dei Messaggi di Posta Elettronica e Newsgroups. Problemi Metodologici e Tecnici." Paper Presented at the Conference AIS "Lo Studio del Mutamento Sociale: Problemi Metodologici e Ricerche empiriche", January 28th–29th.

Herman, E. S. and N. Chomsky. 1988. *Manufacturing Consent: The Political Economy of the Mass Media*. New York: Pantheon.

ISTAT. 1998. *I lettori di libri in Italia*. Roma: ISTAT.

Katz, E. and P. Lazarsfeld. 1955. *Personal Influence*. New York: The Free Press.

Katz, E. 1957. "The Two-Step Flow of Communication, an Up-to-Date Report on an Hypothesis." *Public Opinion Quarterly* 21:61–78.

Kourvetaris, G. A. 1997. *Political Sociology: Structure and Process*. Boston: Allyn & Bacon.

Lanzetti, C. 2002. *Aspetti Metodologici della Ricerca*. (Forthcoming).

Lazarsfeld, P. F., B. Berelson and H. Gaudet. 1944. *The People's Choice: How the Voter Makes Up His Mind in a Presidential Campaign*. New York: Columbia University Press.

Liebes, T. and E. Katz. 1988. "Dallas and Genesis: Primordiality and Seriality in Popular Culture." In *Media, Myths, and Narratives. Television and the Press*, edited by J. W. Carey. London: Sage.

Livolsi, M., ed. 1992. *Il pubblico dei media*. Firenze: La Nuova Italia.

Losito, G. 1994. *Il potere dei media*. Firenze: NIS.

Luhmann, N. 1996. *Die Realität der Massenmedien*. Opladen: West Deutscher Verlag.

Magrini, D. 2002. *Sbatti il web in prima pagina*. Milano: Angeli.

Mante-Meijer, E. *et al.* 2001. "ICT Uses in Everyday Life: Checking It Out with the People — ICT Markets and Users in Europe." Confidential EURESCOM P903 Project Report, EDIN0161-0903.

McQuail, D. 1992. *Media Performance: Mass Communication and the Public Internet.* London: Sage.

Meyrowitz, J. 2003. "Canonic Anti-Text: Marshall McLuhan's Understanding Media." In *Canonic Text in Media Research. Are There Any? Should There Be? How About These?* Edited by E. Katz *et al.* Cambridge: Polity Press.

Meyrowitz, J. 2004. "Mainstream Media Mythologies." *Aprile* 119:12–13.

Minnini, G. and R. Ghiglione. 1995. *La Comunicazione Finzionante. Lo, La Televisione.* Milano: Angeli.

Moores, S. 1993. *Interpreting Audiences. The Ethnography of Media Consumption.* London: Sage.

Neuwirth, K. and E. Frederick. 2004. "Peer and Social Influence on Opinion Expression. Combining the Theories of Planned Behavior and the Spiral of Silence." *Communication Research* 31(6):669–703.

Noelle-Neumann, E. 1984. *The Spiral of Silence. Public Opinion — Our Social Skin.* Chicago: University of Chicago.

Noelle-Neumann, E. 1991. "The Theory of Public Opinion: The Concept of the Spiral of Silence." In *Communication Yearbook 14*, edited by J. A. Anderson. Newbury Park, CA: Sage.

Pertierra, R. 2004. "Globalism, Culture, and the Nation-State." *Philippine Studies* 52(1):119–130.

Popper, K. R. 1994. "Una Patente per Fare TV." In *Cattiva Maestra Televisione*, edited by K. R. Popper and J. Condry. Milano: Reset.

Pucci, E. ed. 2004. *L'industria della Comunicazione in Italia.* Milano: Guerini.

Silverstone, R. 1994. *Television and Everyday Life.* London: Routledge.

Strassoldo, R. 2005. *Gli Atteggiamenti dei Giovani Italiani verso il Mondo Militare.* Soveria Mannelli: Rubettino.

Thompson, J. B. 1995. *The Media and Modernity. A Social Theory of the Media.* Cambridge: Polity Press.

7 Teenagers and Mobile Phones in Malta: A Sociolinguistic Profile

Lydia Sciriha

Introduction

In modern society, telephone communication has come to permeate everyday life. The mobile phone, in particular, has become a ubiquitous feature of contemporary living, deeply affecting the lives of all who have access to it.

The impact that the mobile phone is having on teenagers has recently been the focus of extensive research (e.g., Gillard *et al.* 1996; Manceron 1997; Ling 1999, 2000, 2001; Taylor and Harper 2000; Green 2001; Eldridge and Grinter 2001). Teenagers constitute a special group of mobile users since statistics in most countries reveal that mobile companies are witnessing heavy penetration levels in this segment of the market.

The mobile phone is viewed by teenagers as an indicator that they are growing up and that they are free. Owning a mobile phone marks a transition between childhood and adulthood. It seems that by purchasing mobile phones for their teenagers, parents "release them little by little" (Ling 1999) since they allow their teenage offspring to go out alone, but at the same time, remain under parental surveillance through the mobile (Green 2001). Concomitantly, for most teenagers, having a mobile phone removes them from the threat of social exclusion since most of their friends own one and it has become a common practice to give teenagers mobile phones as Christmas or birthday gifts. For many teenagers, the mobile phone has also become a fashion statement. For this reason, they beg their parents to buy them the latest model, in the same way as they twist their parents' arms to ensure that they are only bought clothes that carry well-known brand names. Of course, there is a price to pay when purchasing a teenager an expensive mobile. Research reveals that mobile phones are the target of criminal acts since quite a number of phones are stolen every day. In fact, Green homes in on this point when she discusses school bans on mobile phones: "The school ban on phones gives institutional recognition to the fact that younger children perceive risk in carrying phones, as they might be stolen from them to support a school-based black market for consumer goods" (2001:39).

Teenagers are at a period in life in which they consider their peers as possibly the most important people. Thus, they spend a lot of the time with them and very often value their ethos much more than that of their parents.

Case Study: The Mediterranean Island of Malta

This paper sifts the most significant findings on teenagers and mobile phones, derived from a large-scale scientifically representative survey (Sciriha 2004b) on the impact of telephony in Malta, a tiny island covering 317.2 square kilometers that is situated at the center of the Mediterranean, 93 kilometers south of Sicily and 288 kilometers north of Africa. The population of Malta stands at around 400,000.

Malta has had a long and chequered history. It was ruled by several colonizers, namely, the Arabs (870–1090); the Normans (1090–1266); the Angevins (1266–1283); the Aragonese (1283–1410); the Castilians (1412–1530); the Order of St. John (1530–1798); the French (1798–1800), and the British (1800–1964). Malta obtained its Independence from Britain in 1964 and is now a member of the European Union.

Maltese, which is the national language of Malta, is also co-official with English. In fact, Section 5 of the Maltese Constitution states that "Every law will be enacted in Maltese and English and in the eventuality of conflict between Maltese and English texts, the Maltese text shall prevail." On the May 8th, 2002. Maltese was accorded the status of one of the official languages of the European Union (Sciriha 2004a).

Quantitative Study on Teenagers and Mobile Telephony

For the purposes of this paper, teenagers are those who are between 14–17 years of age. The main reason for this lies in the fact that they are still legally minors and thus still under parental responsibility.

Is there a high density of mobile phone ownership among Maltese teenagers, and if so, how is the mobile phone changing their lives? Has the mobile phone become important to them? What are the effects of mobile phone ownership among teenagers? Are parents aware of the possible pitfalls when they buy their teenage son or daughter a mobile phone? Why are SMS messages so popular among teenagers?

Methodology of Research

A representative sample of 500 randomly selected respondents from the different towns and villages of Malta and Gozo were interviewed by a team of bilingual interviewers.

A two-stage probability sampling technique was used to interview persons aged 14 years and over. The towns and villages of Malta were first divided into 20 blocks and 500 respondents were selected from these blocks using the standard system of regular intervals, after a random starting point. For each block, the interviewers were assigned a given quota, based on the ideal sampling distribution, in such a way that the quota for each block systematically represented the demographic profile of the Maltese population.

The quota for each block was stratified by gender and age group. At the 95% confidence level, and assuming an anticipated split of 90/10, the sampling error for the total quota is estimated at +/− 2.9%.

The methodology adopted makes the study fully representative of the population of Malta and of its sister island, Gozo. The data is valid for these two communities and the sampling error refers to them.

The instrument used for the survey was a structured questionnaire that was divided into two parts. The first section of the questionnaire consisted of questions pertaining to the interviewee's gender, occupation, level of education, and age, while the second section comprised questions regarding the use of "fixed line" and "mobile" phones.

In view of the linguistic situation in Malta as described above, the questionnaire was produced in both Maltese and English, thus allowing all person-to-person interviews to be conducted in the respondent's native language.

Results

Table 1 shows that of the total sample of 500 persons who took part, only 137 do not own a mobile. It is apposite to highlight the fact that the youngest Maltese rank second in mobile phone ownership, and are overtaken only by persons in the 26–35 age cohort by a mere 0.8% (14–17: 89.2% vs. 26–35: 90%). On the other hand, the respondents in the two oldest age groups are marked by relatively low percentages of mobile phone ownership (51–65: 56% vs.

Table 1. Mobile Phone Ownership

	Mobile	No Mobile	Total
N =	363	137	500
	%	%	%
Age			
14–17	89.2	10.8	100
18–25	87.1	12.9	100
26–35	90.0	10.0	100
36–50	77.3	22.7	100
51–65	56.0	44.0	100
65 +	38.0	62.0	100

Base = All

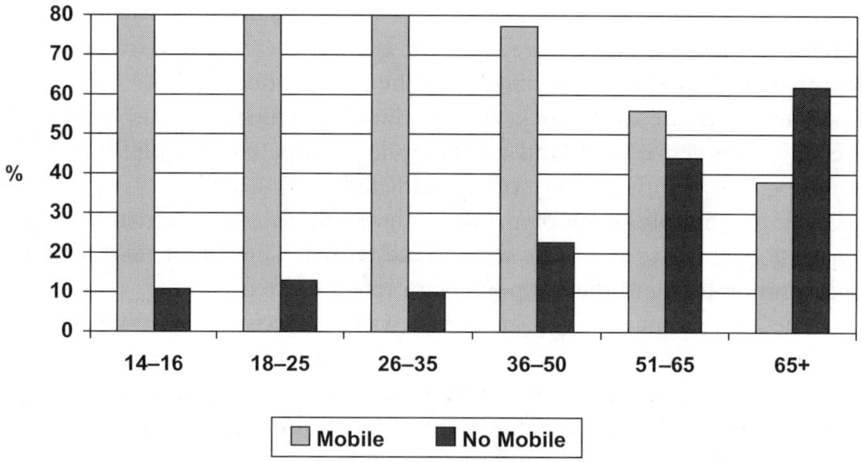

Figure 1. Mobile Ownership

65+: 62%). Figure 1 graphically portrays these differences in mobile phone ownership.

Reasons Why Teenagers Own a Mobile Phone

As is clearly evident in Table 1 and Figure 1, mobile phone usage has penetrated quite strongly in the teenage segment of the market. Teenagers were asked to give reasons why they own a mobile phone. Table 2 gives the results.

When compared with the reasons given by persons in the older age groups, the teenagers' reasons are quite notable. For example, they cited the fact

Table 2. Reasons Why the Respondents Owned a Mobile Phone

	Total	Age					
		14–17	18–25	26–35	36–50	51–65	65+
N=	363	33	61	76	119	47	27
	%	%	%	%	%	%	%
No Answer	0.3	—	—	1.3	0	0	0
Everyone has a mobile	7.4	18.2	8.2	10.5	4.2	2.1	7.4
To be contactable everywhere	70.2	75.8	70.5	69.7	70.6	74.5	55.6
To be free, no need to be tied to one place	13.5	9.1	18.0	14.5	10.9	17.0	11.1
To feel secure	28.9	12.1	32.8	26.3	31.1	27.7	40.7
Status symbol	1.9	3.0	3.3	3.9	0.8	0.0	0.0
Necessity	9.9	6.1	8.2	10.5	10.1	10.6	14.8
For emergencies	2.8	3.0	3.3	2.6	2.5	2.1	3.7
A gift	5.0	—	4.9	1.3	5.9	14.9	—
For work	3.3	—	—	7.9	5.0	—	—
Contact children	1.1	—	—	1.3	2.5	—	—

Base = All

that "everyone has a mobile phone" (14–17: 18.2%), the most of all the respondents in the different age cohorts (18–25: 8.2%; 26–35: 10.5%; 36–50: 4.2%; 51–65: 2.1% and 65+: 7.4%). This is symptomatic of their desire to be like their peers. Moreover, they also gave as a reason the fact that they can be "contactable everywhere" (18–50: 75.8% vs. 18–25: 70.5%; 26–35: 69.7%; 36–50: 70.6%; 51–65: 74.5%; 65+: 55.6). Such statistics are in line with studies that reveal that one of the salient reasons parents buy their teenage children a mobile phone is to be able to track them down, and teenagers are all too cognizant of this fact. Interestingly, however, the teenagers were the ones who gave the reason "to be free, no need to be tied down to one place" the least mention from among the age groups, because in reality, they are not really free since their parents monitor their moves by means of the mobile.

Do Teenagers Enjoy Being Phoned?

All of those taking part in the survey, regardless of whether they owned a mobile phone or not, were asked whether they enjoyed being phoned.

Table 3 and Figure 2 reveal that teenagers are the most enthusiastic about being phoned. In fact, 89.2% of them said that they enjoy being phoned, while lower percentages were registered by respondents in the other age groups, especially those in the 36–50 age cohort (70.1%).

Table 3. Whether the Respondents Enjoyed Receiving Telephone Calls

	Total	Age Groups					
		14–17	18–25	26–35	36–50	51–65	65+
N=	500	37	70	84	154	84	71
	%	%	%	%	%	%	%
Yes	77.6	89.2	87.1	76.2	70.1	75.0	83.1
No	22.4	10.8	12.9	23.8	29.9	25.0	16.9

Base = All

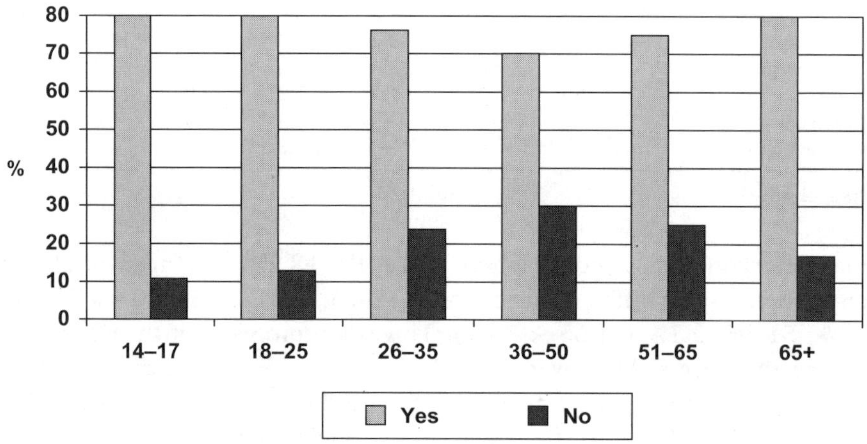

Figure 2. Whether the Respondents Enjoyed Receiving Phone Calls

The respondents who enjoyed being phoned were also asked to give reasons for this. Table 4 and Figure 3, which give the reasons by age group, reveal that teenagers are particularly happy when individuals, undoubtedly other teenagers like themselves, phone them because that shows that "people think of me" (69.7%). More of them gave this reason than those in the other age groups. Another reason the Maltese gave is that when they speak on the phone they "don't feel lonely." While the oldest persons were the ones who cited this reason most frequently (30.5%), they were closely followed by the teenagers (27.3%). Another popular reason given was that "speaking on the phone relaxes me." Quite noteworthy is the fact that this reason was given the most by the two youngest age groups (14–17: 21.1% vs. 18–25: 26.2%).

Table 4. Reasons Why the Respondents Enjoyed Being Phoned

	Total	Age					
		14–17	18–25	26–35	36–50	51–65	65+
N=	388	33	61	64	108	63	59
	%	%	%	%	%	%	%
People think of me	64.2	69.7	62.3	67.2	63.9	65.1	59.3
Don't feel lonely	24.5	27.3	23.0	21.9	23.1	23.8	30.5
Speaking on the phone relaxes me	14.2	21.2	26.2	12.5	9.3	9.5	13.6
Speak to friends and relatives	3.4	—	3.3	4.7	3.7	3.2	3.4

Base = All

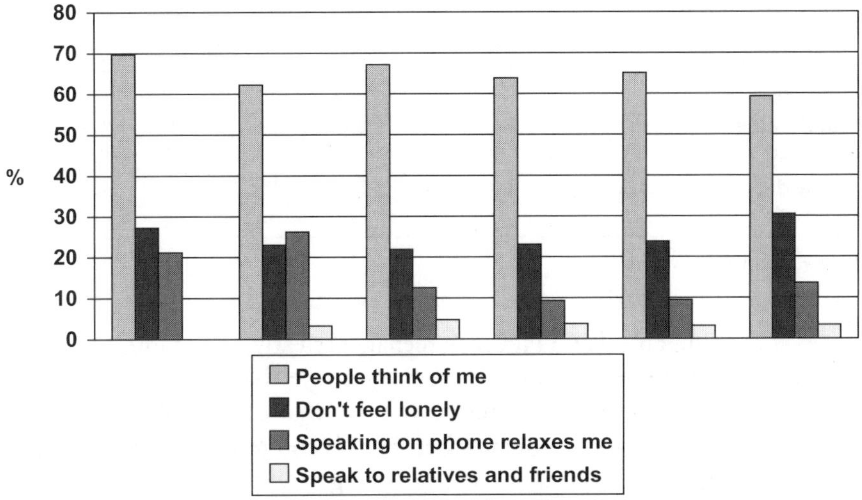

Figure 3. Reasons Why the Respondents Enjoyed Being Phoned

Why Do Teenagers Phone Using a Mobile?

Teenagers enjoy communicating on the phone because phoning others relaxes them and keeps them in touch with what is happening in their peer groups. Teenagers consider their peers to be significant others. Since the overwhelming majority of the teenagers (89.2%) in the present survey owned a mobile, the teenage mobile owners were also asked to give reasons why they would incur more expenses to phone others on their mobile phone.

Table 5. Reasons Why the Respondents Phoned a Mobile Line

	Total %	14–17 %	18–25 %	26–35 %	36–50 %	51–65 %	65+ %
N=	363	33	61	76	119	47	27
No Answer	2.8	6.1	3.3	3.9	0.8	4.3	0.0
So that only the addressee knows that I am calling	6.9	6.1	4.9	10.5	5.0	6.4	11.1
Only the person I call will answer the phone	31.4	39.4	45.9	27.6	30.3	25.5	14.8
Don't waste time phoning other numbers	21.2	9.1	27.9	22.4	17.6	19.1	37.0
I don't need to wait on hold to speak to the addressee	16.8	18.2	16.4	15.8	19.3	12.8	14.8
To make urgent calls	14.9	9.1	11.5	7.9	20.2	19.1	18.5
So that only the addressee knows I am calling	6.9	6.1	4.9	10.5	5.0	6.4	11.1
When the addressee is outdoors	11.8	3.0	9.8	15.8	13.4	12.8	7.4
Will definitely find the addressee	7.7	6.1	6.6	9.2	8.4	8.5	3.7
I want the conversation to be private	5.8	12.1	3.3	3.9	5.9	4.3	11.1

Base = All mobile owners

As shown in Table 5 and Figure 4, teenagers clearly do not mind spending more money to phone a mobile number to ensure that their privacy is safeguarded. Gillard *et al.* (1996) stressed the fact that the mobile phone gives teenagers greater privacy to contact their friends and to be contacted by their friends, without their parents' knowledge (Ling 1998), since teenagers are at the stage in their life when they want to break free of their parents' invisible leash.

Having a mobile phone means that teenagers can interact with their peers without their parents' knowledge. In fact, the teenagers in our survey valued most the fact that "only the person I call will answer the phone" (39.4%), which effectively means that they now bypass the traditional gatekeepers at home, namely the parents. Recent research (Kopomaa 2000; Taylor and Harper 2001) has shown that because teenagers are now communicating directly by phoning each other on the mobile, they are in fact sidestepping their parents and their parents no longer know with whom their children hang around. Thus, for teenagers, the mobile phone has provided them with the opportunity to make friends with anyone, even with those who would be

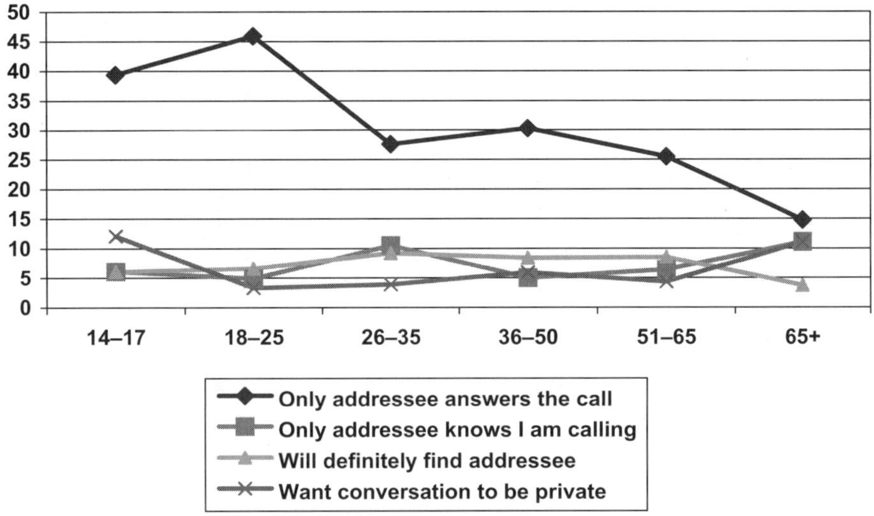

Figure 4. Reasons Why the Respondents Phoned on a Mobile Number

considered by their parents to be "undesirables" or a "bad influence." Furthermore, teenagers want to have a private life, away from persistent questioning by their parents as to "who is calling?," "what does s/he want?," "why is s/he your friend?" and other questions that teenagers consider highly reminiscent of the Spanish Inquisition and thus a flagrant invasion of their privacy. Thus, unsurprisingly, teenagers were the ones who cited the reason that they want "the conversation to be private" (12.1%) the most, when compared to the other respondents in the other age groups (18–25: 3.3%; 26–35: 3.9%; 36–50: 5.9%; 51-65: 4.3%; 65+: 11.1%).

The reasons given by the teenagers all point to a gnawing concern of theirs, namely their private life. Funston and MacNeill (1999), in their quantitative survey of mobile phone usage in Australia, discovered that "caller privacy" was one of the reasons why teenagers bought a mobile phone. It is evident from the relatively high percentages registered by teenagers when compared with the other respondents that having a private life has now become easier with the advent of the mobile phone.

Mobile Phone Traffic

As regards the duration of calls made to mobile numbers by mobile and non-mobile owners, the results revealed that a total of 41.5% of mobile owners do

Table 6. Duration of Mobile Calls

	Total	14–17	18–25	26–35	36–50	51–65	65+
N=	363	33	61	76	119	47	27
	%	%	%	%	%	%	%
No Answer	0	—	—	—	—	—	—
0–15 seconds	11.6	6.1	6.6	5.2	15.1	21.3	14.8
16–30 seconds	13.5	21.2	11.5	15.8	13.5	12.8	3.7
46–60 seconds	16.8	18.2	14.8	21.1	15.1	19.1	11.2
1–2 minutes	30.9	21.2	32.8	30.3	31.2	27.7	44.4
2–3 minutes	10.7	21.2	6.5	11.8	9.2	8.5	14.8
3–4 minutes	3.3	—	6.5	2.6	3.4	2.1	3.7
4–5 minutes	7.7	3.0	14.8	6.6	9.2	2.1	3.7
5–10 minutes	4.1	3.0	6.5	5.3	2.5	4.3	3.7
11–20 minutes	0.8	6.1	—	—	0.8	—	—
30+ minutes	0.6	—	—	1.3	—	2.1	—

Base = All mobile owners

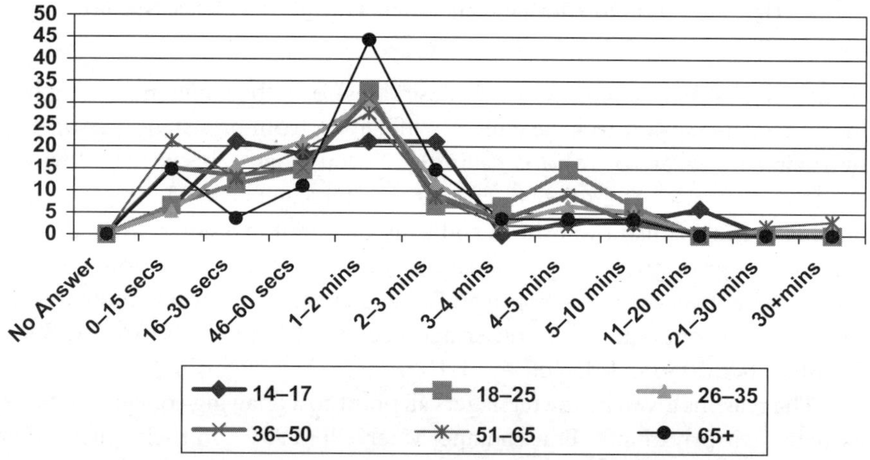

Figure 5. Average Duration of Calls by Mobile Phone Owners

not exceed 60 seconds when they phone on their mobile and 30.9% of the mobile owners do not talk on their mobile for more than 2 minutes.

Table 6 and Figure 5 show that the teenagers' calls on their mobiles on average do not exceed 16–60 seconds. However, of the 10.7% of persons whose calls are between 2–3 minutes, teenagers register the highest percentage of all the other age groups. Besides, it is particularly noteworthy that 0.8%

of the calls are on average between 11–20 minutes, and that the highest percentage of such calls were made by teenagers (14–17: 6.1%) and respondents in the 36–50 age cohort (0.8%).

Who Sends the Most SMS Messages?

The sending of SMS messages is a phenomenon that has taken mobile companies by surprise. The traffic statistics of SMS messages reveal that it is the younger respondents who use this relatively cheap service the most. In fact, as shown in Table 7 and Figure 6, 32% of mobile phone owners said that they send over 20 SMS messages every week. Moreover, of these, the highest percentage of those who said that they send so many SMS messages belongs to the youngest age group. In fact, a staggering 78.8% of the teenagers said that they send over 20 SMS messages every week as compared to 10.6% of those who are between 50–65 years of age.

Who Receives the Most SMS Messages?

SMS messages resemble the ordinary exchange of e-mail messages, where one usually expects a reply to a message sent. The respondents were accordingly asked about the average number of SMS messages they receive per week. Table 8 and Figure 7 show that many respondents received more SMS messages than they actually sent. Some of the messages that they receive are probably instances of what Malinowski calls "phatic communion" (1923:315), in which individuals exchange words to convey their links with one another. These words do not have meaning per se but "fulfill a social function and that is their principal aim" as in, for example, an SMS message such as "It's hot today!"

Table 7. Average Number of SMS Messages Sent Every Week

	Total	14–17	18–25	26–35	36–50	51–65	65 +
N =	363	33	61	76	119	47	27
	%	%	%	%	%	%	%
None	29.2	3.0	6.5	14.5	38.7	51.1	74.1
1–5	12.7	3.0	3.3	19.7	18.5	6.3	11.1
6–10	10.2	—	8.2	14.5	11.7	14.9	—
11–15	7.4	6.1	3.3	9.2	8.4	12.8	—
16–20	8.5	9.1	9.8	9.2	10.9	4.3	—
Over 20	32.0	78.8	68.9	32.9	11.8	10.6	14.8

Base = All mobile users

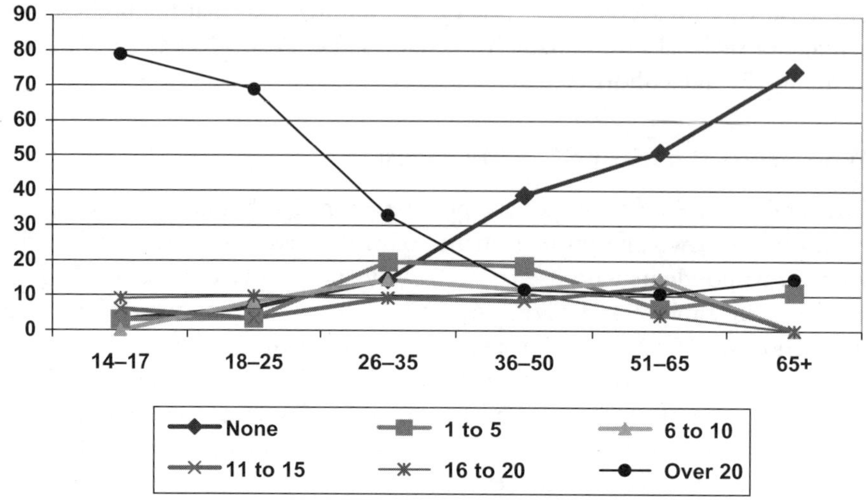

Figure 6. SMS Messages Sent Every Week

Table 8. Average Number of SMS Messages Received Every Week

	Total	Age					
		14–17	18–25	26–35	36–50	51–65	65+
N=	363	33	61	76	119	47	27
	%	%	%	%	%	%	%
None	21.2	3.0	1.6	9.2	29.4	40.4	51.9
1–5	16.0	0	4.9	21.1	21.1	12.8	29.6
6–10	11.3	3.0	8.2	10.5	15.1	17.0	3.7
11–15	8.3	6.1	8.2	10.5	8.4	10.6	—
16–20	9.6	12.1	11.5	10.5	10.9	6.4	—
Over 20	33.6	75.8	65.6	38.2	15.1	12.8	14.8

Base = All mobile phone owners

Once again, it is the teenagers in the youngest age group who receive the most SMS messages (over 20 messages: 75.8%), followed by the second youngest age group (18–25: 65.6%). The oldest participants receive more SMS messages than those in the 51–65 age cohort (65+: 14.8% vs. 51–65: 12.8%).

Reasons Why the Respondents Send SMS Messages

As previously shown in Table 5 and Figure 4, the reasons teenagers gave for phoning others on a mobile instead of on a fixed line phone is because the

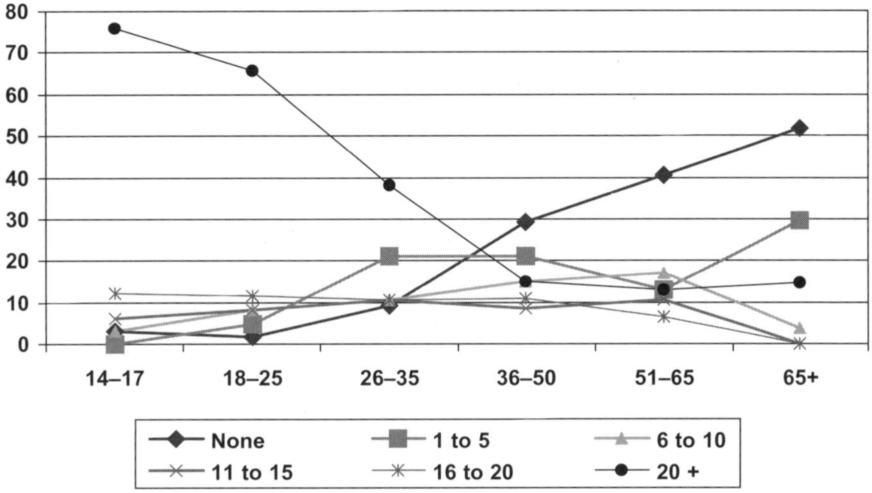

Figure 7. Average Number of SMS Messages Received Every Week

mobile phone affords them privacy in that no one, other than the addressee, needs to know who is calling whom.

Sending SMS messages is also a predominantly teenage method of communicating with others. According to Fox, "Texting is particularly important in maintaining contact with a wide social network. Texting helps teenagers to overcome awkwardness and inhibitions and to develop social and communicative skills — they communicate with more people, and more frequently, than they did before mobiles" (2001:2). For this reason, the participants were asked to give reasons why they send SMS messages. Table 9 and Figure 8 give a breakdown of the results by age. Teenagers send SMS messages for a number of reasons but mainly to "communicate with family and friends" (27.3%), to communicate with someone privately (12.1%), to "obtain information about an appointment" (24.2%), because "SMS is cheaper than calling on the mobile," (63.6%) and also because sending an SMS message is "fun" (18.2%) or to "kill time" (3.0%). As is evident in Table 9, relatively more teenagers than members of the other age cohorts give these reasons.

Quite interesting is the fact that the reasons given by teenagers for sending SMS messages are very similar to those they gave when asked why they phone people on their mobile phones. It is strikingly clear that teenagers are moving towards independence more quickly than before. They can schedule meetings with their friends without their parents' direct knowledge. A simple SMS

Table 9. Reasons Why the Respondents Send SMS Messages

	Total	Age					
		14–17	18–25	26–35	36–50	51–65	65+
N=	363	33	61	76	119	47	27
	%	%	%	%	%	%	%
NA	24.8	3.0	4.9	10.5	34.5	42.6	63.0
To maintain contact with family and friends	19.6	27.3	26.2	27.6	16.0	6.4	11.1
To inform/obtain information about appointments	10.2	24.2	14.8	11.8	8.4	2.1	—
SMS is cheaper than calling on a mobile	43.3	63.6	55.7	46.1	35.3	36.2	29.6
Fear of intruding by phoning	8.3	3.0	19.7	3.9	8.4	8.5	—
To communicate with someone privately	9.9	12.1	16.4	14.5	7.5	4.3	—
It is fun	8.3	18.2	18.0	9.2	4.2	2.1	—
Send SMS when messages are not urgent	1.1	3.0	1.6	1.3	—	2.1	—
Sending an SMS is easier and quicker	1.1	—	1.6	—	1.7	2.1	—
Send an SMS to answer an SMS received	0.8	—	—	2.6	0.8	—	—
For emergencies	0.6	3.0	—	1.3	—	—	—
To kill time	0.6	3.0	1.6	—	—	—	—

Base = All mobile owners

message provides all of the details and the receipt of such a message is also very discreet. Teenagers can put their phone on silent mode and still know that they have a message without their parents' knowledge.

One of the reasons that was given, and that does not feature in the table listing the reasons for phoning on a mobile, is the fact that "sending an SMS message is cheaper than phoning." One would think that teenagers are the most price sensitive of all the age groups in this regard. Although most of the teenagers did not exceed 60 seconds of talk time, quite surprising is the fact that there was a very small percentage who said that they talk for up to 20 minutes. Since most teenagers use pre-paid cards and tend to pay for their mobile phone expenses out of their pocket money, they might

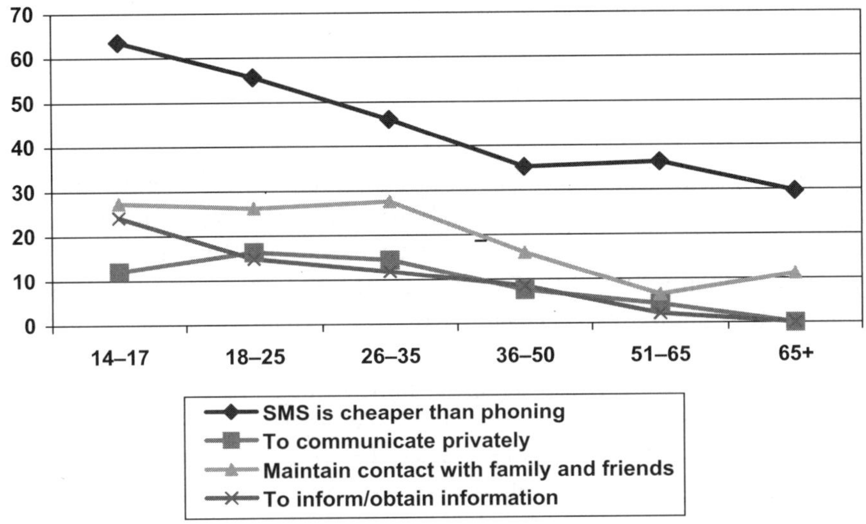

Figure 8. Reasons Why the Respondents Sent SMS Messages

not really be in a position to talk on the phone since every second costs money. But it seems that Maltese teenagers prefer to talk on the phone rather than to buy other things such as cigarettes and chocolates out of their pocket money.

The effects of having a mobile phone is positive when considering a study by Bates and Charlton, who have hypothesized that the drop in smoking among 15 year olds in the United Kingdom (from 30% to 23% between 1996–1999) is linked to the limited financial resources of teenagers. They prefer to spend their pocket money on topping up their phone card rather than buying cigarettes. For teenagers, having a mobile phone offers the same teenage needs as smoking does. Having a mobile phone "offers an adult lifestyle, sociability, rebellion and peer group bonding. For teenagers, smoking may be seen as 'old technology' with the bright new world of text messaging, email and WAP becoming the new aspirational gateway to adult life" (2000).

In a somewhat different vein, an article by the BBC entitled "Mobile phones cause choc sales to fall" blames mobile phones for the decline in chocolate sales in the U.K. The reason for this decline is similar to the decline in smoking among teenagers. They prefer buying top-up cards to chocolates. But both studies reveal that teenagers have to budget their weekly pocket money now that they have to top up their mobile phone and that, in doing so, they have had to give up some of their previous "luxuries" such as cigarettes or chocolates.

Where are You?

Research on teenagers and mobile phone use reveals that some parents purchase their teenage children a mobile phone to keep tabs on them, to monitor their moves (Green 2001), or rather what Raskow and Navarro have dubbed "remote mothering" (1993), namely keeping children on an invisible leash. Parents often phone their teenage children on the mobile to keep in touch with them if they are out on the weekends, and often the first question parents ask is "Where are you?" At times, some teenagers are economical with the truth and blatantly lie to their parents when they are asked that question. Qualitative research by the present writer has revealed that it is becoming a common practice to lie to callers who ask "Where are you?," especially if the callers are their parents. It would not the first time that teenagers told their parents that they were enjoying themselves in an entertainment resort in Malta, when in truth, they were on the sister island of Gozo! Most parents now seem relieved that their teenagers have mobiles because they know where they are. But, of course, although it is to be hoped most teenagers are truthful, a survey by OmniPoint in Northeastern United States revealed that around 20% of the respondents said that they lie when they are asked "Where are you?" (Chihara 2000), since it is much easier to lie on the phone than face to face. Furthermore, teenagers whose parents are prone to asking the dreaded question "Where are you?" might circumvent the receipt of calls and being asked questions, by switching off their telephone or by saying that they did not hear the phone ring, especially if they were enjoying themselves at a disco. Qualitative research has shown that this is one of the most effective ways of barring callers, especially parents. In fact, when teenagers are asked by their ever-anxious parents why they did not answer the phone, they would say that they had "a flat battery." Roos has dubbed this popular way of circumventing unwelcome questions the "dead battery syndrome" (2001).

Furthermore, it is surprising that, as is evident in Table 10 and Figure 9, only 28.1% of the respondents said that they not happy to be asked "where are you" by callers. However, quite remarkable is the fact that of those who said that such a question did not bother them, the youngest participants registered the lowest percentage, with teenagers the lowest of all (14–17: 66.7% vs. 65+: 81.5%). Many feel that being asked to give their location reveals inquisitive callers and it is particularly significant that the youngest two age

Table 10. **Whether the Respondents Were Annoyed when Callers Asked "Where Are You?"**

	Total	Age					
		14–17	18–25	26–35	36–50	51–65	65+
N=	363	33	61	76	119	47	27
	%	%	%	%	%	%	%
No, it does not annoy me	74.9	66.7	67.2	80.3	75.6	76.6	81.5
Yes, it does. I feel that the callers are inquisitive	21.8	27.3	29.5	17.1	21.0	21.3	14.8
Yes, it does. Such information is irrelevant	6.3	6.1	8.2	5.3	5.9	8.5	3.7

Base = All mobile owners

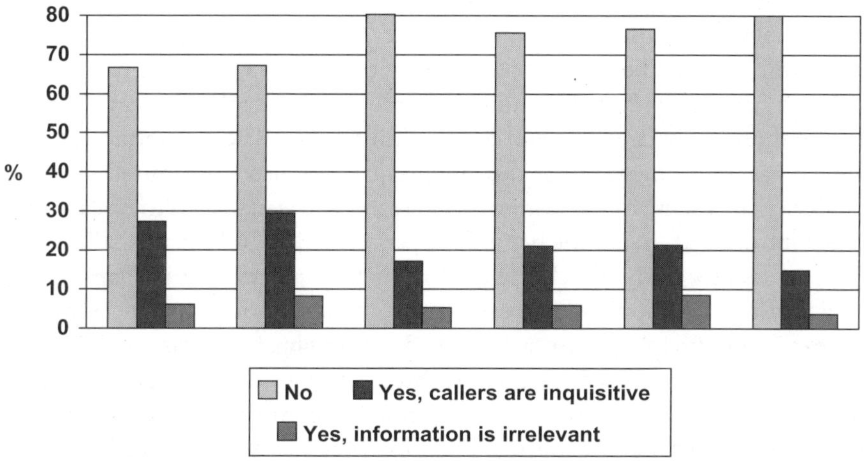

Figure 9. **Whether the Respondents Were Annoyed When Asked "Where Are You?"**

group members aired their annoyance more intensely than the other age groups.

From the results of the survey, which focused on teenagers' use of the mobile phone, it is evident that teenagers are the most enthusiastic users of mobile phones. The mobile has provided them with an earlier-than-anticipated independence from their parents, since the latter are all too happy and relieved that their children are fine as long as they carry the mobile phone. Of course, like so many other objects in life, there are both advantages and disadvantages to having a mobile. A mobile phone provides safety and security to teenagers in emergencies (Ling 1998) when they are out of the house, but can be a loaded weapon, if not used responsibly. The popularity of sending SMS messages is linked with their wanting independence and no interference from their parents. Nowadays, parents no longer know who their children's friends really are, since teenagers keep those who would be considered "undesirables" by their parents on their mobile phone telephone directory and privately interact with them by voice mail or through SMS messages.

The mobile phone has provided a new means of caring for the young and the elderly alike: many parents buy their teenage children a mobile phone because they are comforted by the thought that they can reach their children by mobile. The feeling of immediate contact can be real enough, but the sense of security can also be unreal. Already, research is showing that children lie to their parents when contacted on their mobile.

This study has provided ample data to show that the telephone, especially mobile telephony, affords privacy to its user.

All in all, it can be stated that in the contemporary world, the need to stay in touch has acquired new prominence, and mobile telephony is providing an easy and relatively cheap way of accomplishing this. The telephone can, and indeed has, saved lives, but at the same time it has also contributed to giving modern man a sense of freedom when in reality it has managed to enslave him. Once introduced, it cannot be ignored, and the demand for it is still on the increase, as this study has clearly shown. As time passes, and perhaps as man realizes that the trappings it creates are sometimes not as beneficial as they appear, more disciplined use will be made of it. This is the case with TV in a number of advanced societies, in which the number of hours people expose themselves to has decreased and more time is spent on other activities. But that requires a change in culture, a process which is still incipient in the case of Malta.

References

Bates, C. and A. Charlton. 2003. "Are Mobile Phones Reducing Teenage Smoking?" Accessed August 20th 2003. http//www.globaltechnoscan.com.

Brown, B., N. Green and R. Harper, eds. 2001. *Wireless World: Social and Interactive Aspects of the Mobile Age*. London: Springer.

Chihara, M. 2000. "Lying on the Go." *Boston Phoenix*, March 18th.

Eldrige, M. and R. Grinter. 2001. "Studying Text Messaging in Teenagers." Position Paper for CH1 Workshop No 1: Mobile Communications: Understanding Users, Adoption and Design.

Fox, K. 2001. "Evolution, Alienation and Gossip: The Role of Mobile Telecommunications in the 21st Century." http://www.sirc.org/publik/gossip.shtml.

Funston, A. and K. MacNeill. 1999. "Mobile Matters: Young People and Mobile Phones." Victoria, Australia: Communications Law Centre.

Gillard, P., A. Bow and K. Wale. 1995. *Privacy and Control: Social Indicators of Interest in Future Telecommunications.* Telecommunications Needs Research Group. Melbourne: Royal Melbourne Institute of Technology. http://www.sims.monash.edu.au/research/itnr/papers/privcon.html.

Green, N. 2001. "Who's Watching Whom? Monitoring and Accountability in Mobile Relations." In *Wireless World: Social and Interactive Aspects of the Mobile Age*, edited by B. Brown, N. Green and R. Harper. London: Springer.

Kopomaa, T. 2000. *The City in Your Pocket — Birth of the Mobile Information Society.* Tampere, Finland: Gaudeamus.

Ling, R. 1998. "'We Will Be Reached': The Use of Mobile Telephony Among Norwegian Youth." Kjeller, Telenor Forskning og Utvikling, FoU Rapport 16.

Ling, R. 1999. "We Release Them Little by Little: Maturation and Gender Identity As Seen in the Use of Mobile Telephone." International Symposium on Technology and Society (ISTAS '99) Women in Technology and Professional Perspectives, July 29th–31th, at Rutgers University, New Brunswick, New Jersey.

Ling, R. 2000. "Norwegian Teens, Mobile Telephony and SMS-Use in School." Telenor Forskning og Utvikling, FoU Rapport 7.

Ling, R. 2001. "The Diffusion of Mobile Telephony Among Norwegian Teens: A Report from After the Revolution." Paper Presented at ICUST 2001, June, at Paris, France.

Ling, R. 2002. "The Social Juxtaposition of Mobile Telephone Conversations and Public Spaces." Paper Presented at the Conference on The Social Consequences of Mobile Telephones, July, at Chunchon, Korea.

Malinowski, B. 1923. "The Problem of Meaning in Primitive Languages." In *The Meaning of Meaning: A Study of the Influence of Language Upon Thought and of the Science of Symbolism*, edited by C. K. Ogden and I. A. Richards. London: Routledge and Kegan Paul.

Manceron, V. 1997. "Get Connected — Social Uses of the Telephone and Modes of Interaction in a Peer Group of Young Parisians." In *Blurring Boundaries: When are Information and Communication Technologies Coming Home?* Edited by A. Kant and E. Mante-Meijer. Stockholm: Norstedts.

Rakow, L. F. and V. Navarro. 1993. "Remote Mothering and the Parallel Shift, Women Meet the Cellular Phone." In *Critical Studies in Mass Communication* 10: 144–157.

Roos, J. P. 2001. "Postmodernity and Mobile Communications." Paper Presented at the ESA Helsiinki Conference, August. http://www.valt.helsinki.fi/staff/proos/mobilezation.htm.

Sciriha, L. 2004a. *Beginning Maltese*. Malta: Malta University Publishers.

Sciriha, L. 2004b. *Keeping in Touch — The Sociolinguistics of Mobile Telephony*. Malta: Agenda Publishers.

Taylor, A. S. and R. Harper. 2001. "Talking 'Activity': Young People & Mobile Phones." Paper Presented at the CHI 2001, Mobile Communications Workshop, Seattle, Washington.

8 *Mobile Phones, Aged Homes, and Family Relations in Hong Kong Preliminary Observations*

William Wai-lim Wong

Introduction

It is generally recognized that the fixed-line telephone transcends time and space and that the territory used to restrict contacts is no longer dominant. The mobile phone being a portable personal communicative device, further embraces this process of de-territorization. It creates a spatial dimension for the parties involved so that the autonomy that people do not usually enjoy with a fixed-line phone can be exercised at will. The mobile phone can provide an important linkage in the organization of daily life (Ling 2004) and the management of emotional labour for its users and their family members. In the case of residential care, the mobile phone can also perform such tasks among family members who are not living under the same roof.

The present inquiry attempts to explore the use of mobile phones among residents of old people's homes. First, it aims to examine the family relationships between the residents of such homes and their family members with regard to the use of the mobile phone. It is significant to know how elderly people who have moved to homes communicate with their family members through the use of mobile phones. It is assumed that elderly people who have moved to homes will miss their families and vice versa. How the mobile phone can sustain and enhance their family ties are the key concerns of this research. Second, communication patterns between residents and their families in the co-construction of the world of the mobile phone and its network will be explored.

This chapter will begin with a brief introduction to the relationship between the fixed-line phone and the mobile phone, and the background of the long-term care policy for old people in Hong Kong. It will then be followed by a description of the findings. The final section will consist of a discussion of the use of mobile phones among users in old-age homes.

The Changing of the Guard: Fixed-line Phones or Mobile Phones

At present a low-technology fixed-line telephone system and a high-technology mobile phone system coexist. It is useful at this point to highlight the features of the fixed-line phone in Hong Kong and its relationship to the mobile phone.

Fixed-line telephones can be found all over Hong Kong in shops, restaurants, and public facilities. For quite a number of years it has almost become a common practice for people in Hong Kong to borrow the use of a fixed-line telephone whenever and wherever they wish. With the exception of using a telephone in a public telephone booth, borrowing the use of a fixed-line telephone in Hong Kong is free of charge; a flat rate is charged for a fixed-line phone on a quarterly basis (about HK$330/four months) (Ming Pao 2004) irrespective of the number of calls made or the duration of the calls. Consequently, lending and borrowing the use of such telephones is a common practice. Commercial places even encourage people to borrow their fixed phones as a service promotion strategy.

Hong Kong has entered an era in which increased mobile phone usage has resulted in the fixed-line phone becoming less easily available. Fixed-line phones are now quite invisible except in tea houses or restaurants. As the respondents of this study indicated, this has made it difficult for older people to borrow fixed-line phones when they are out in the street.

On the other hand, the price of mobile phones has come down from a few thousand dollars to an affordable price of a few hundred dollars. Some operators even give out free phones for users if they commit to a fixed-length contract. Network service fees have also plunged, due to keen competition and popularity. Hong Kong is reported to have the highest mobile penetration in the region at 83%. Ownership of mobile phones begins when children reach the ages of 10–12 years (Asia Business Consulting 2004).

Mobile Phones and the Policy on Old People's Homes

Before proceeding to analysis of the data, it is necessary to briefly describe the development of long-term care policy and facilities in Hong Kong. These contextual elements are significant to the understanding of the present study.

Residential homes for elderly people in Hong Kong consist of both homes of the voluntary sector and private commercial homes. As far back as 1972, the Hong Kong government formed a working party to investigate the increasingly aging population and the lack of provision for them. The working party's report, produced in 1973, recommended "care in the community" as the guiding principle for the care of older people (Working Committee Report on the Future Needs of the Elderly 1973). The aim is to enable the elderly, with the support of services, to remain in the community and in their own familiar places as long as possible (1973:15).

There are many reasons for supporting this policy. The key argument is the belief that it is the most economical solution for the community as a whole, since it makes the least demand on community resources, whether these be finance, skilled manpower or accommodation (1973). In order to assess the effect of the "care in the community" approach on the demand of home places, the working committee recommended a slowdown in the pace of construction of homes for two years, rather than in the production of more residential care places (1973:22). By then, the waiting time for admission to a home could be as long as three to five years, depending on the kind of facilities on offer.

A Green Paper (consultation) on services for the elderly (Hong Kong Government 1977) and a White Paper (policy paper) on development of social welfare (Hong Kong Government 1979) all supported this policy objective. In 1994, the policy was also known as "aging in place" (Working Group on Care for the Elderly 1994:48).

However, over the years, due to the steady aging of the population, and the moratorium on the construction of care homes, the shortage of home places has become acute. Consequently, there has been a rapid increase in private, commercial homes. Some private homes are of poor quality and provide sub-standard facilities, dampening any desire for admission. In order to monitor the process of applications for home places, a waiting list and a central referral system have been established (Social Welfare Department 2005).

Despite all the planning and development of the aging policy, there is still no shortage of queries (Chow 2000; Kwok 2001; Chan and Phillips 2002). Under such circumstances, apart from the limited supply of home places, the family remains the major provider of care for elderly people in Hong Kong because of the inadequate provision of social security and community care facilities (Lee *et al.* 2000).

As indicated by the respondents in this study, most elderly people enter homes for reasons such as health, family relations, or insufficient living space at home. On the whole, people do not usually consider old-age homes as an alternative form of living arrangement, mainly because they fear extreme social exclusion or are reluctant to make changes and adjustments due to unfamiliar conditions. The popularity of the mobile phone is somehow contributing to a better feeling among some old people about being admitted to old-age home.

Homes for Elderly People and Their Residents

According to the Hong Kong 2001 Census, 11.3% of people aged 65 and over are living alone; 75.2% of them live with their families, either with their children (56.8%) or spouse (18.4%); and 8% of them live in residential care (Census and Statistics Department 2001).

Those living in residential care live in four different kinds of homes in Hong Kong (Social Welfare Department 2005). These are classified according to the physical and mental abilities of the residents: hostels, homes, care and attention homes, and nursing homes. *Hostels* cater to elderly people who have good self-care abilities, and organized programmes are provided with round-the-clock staff support. There are four such facilities in Hong Kong, and residents from three of these four hostels were interviewed for the present study. *Homes* for the aged provide residential care, meals and a limited degree of assistance in activities of daily living for elderly people who are unable to live independently in the community. The residents are not dependent on assistance with personal or nursing care. *Care and attention homes* are meant for those elderly people who are in need of personal care and limited nursing care, but who are mentally suitable for communal living. *Nursing homes* aim to provide care to elderly people who cannot be adequately cared for in care and attention homes and yet do not require the service provided in infirmaries. Users suffer from poor health with deficiency in activities of daily living but are mentally suitable for communal living (Social Welfare Department 2005).

Research Method

This is an exploratory study. Homes operated by the voluntary sector were chosen for the survey according to their geographical locations and the physical characteristics of the residents. In Hong Kong, the voluntary sector refers to social service organizations, and the operation of their old-age homes is mainly supported financially by the government. Voluntary sector homes in Hong Kong provide communal living, with four to eight persons in a room, sharing a toilet and bathroom with other residents. The reasons for choosing the voluntary sector as sites of data collection are that the residents mostly have a common socioeconomic background, and have different levels of self-care ability. Voluntary sector homes can be found among the three distinct geographical regions of Hong Kong Island, the Kowloon peninsula, and the new towns of the New Territories. In order that distinctions can be drawn between residents with different self-care abilities and from different geographical locations, hostel and care and attention homes were selected from different parts of Hong Kong: two homes were chosen from Hong Kong, three from Kowloon, and three from the New Territories (see Table 1).

Three homes were hostels, while the remaining five were care and attention homes. Hostels are located inside public housing estates, usually on the first and second floors of the building, and care and attention homes are purpose-built buildings. One of the care and attention homes located in the New Territories is run by a Buddhist organization, which serves vegetarian meals and admits female residents only. The hostel in Hong Kong Island has a high percentage of ex-fishermen because of its location. The types of homes, regional distribution and numbers of residents are shown in Table 2.

Data for the research were collected from a survey based on two sets of interview schedules: one for the home management and the second for residents. The questionnaire for the management contains factual questions such as the size of the homes, the age and gender differences of residents, the number of mobile phone users, and the changes brought about by mobile phone users to the homes as a whole. The questionnaire for the residents is slightly longer, and also aims to obtain factual data. It includes questions on personal particulars and family background. The aim of the questionnaire is to collect data about mobile phone usage, communication patterns, cost, manufacturer, network, and the design of the mobile phones used by the residents.

Thirty-three mobile phone users living in the eight homes were interviewed, with an equal number of residents from homes in Hong Kong, Kowloon, and the New Territories, respectively. Of these, 8 were men (constituting 26% of male mobile phone users) and 25 were women (constituting

Table 1. Regional Distribution and Types of Home (N = 8)

Region	Hostels	Care and attention homes	Total
Hong Kong	1	1	2
Kowloon	1	2	3
New Territories	1	2	3
Total	3	5	8

Table 2. Types of Homes, Regional Distribution and Number of Resident

	Care and attention			
Types of homes	Hostels	Homes		Total
No. of homes	3	3	2	8
Regional distribution				
Hong Kong (residents)	148	183	—	331
Kowloon (residents)	126	143	246	515
New territories (residents)	139	148	320	607
Total residents	413	474	566	1453

Table 3. Sample of Mobile Phone Users Interviewed

Age	Male			Female			Total		
	Sample interviewed	Total number of users	(%)	Sample interviewed	Total number of users	(%)	Sample interviewed	Total number of users	(%)
60–65	0	1	0	0	1	0	0	2	0
66–70	0	3	0	3	10	30	3	13	23
71–75	1	8	13	8	15	53	9	23	39
76–80	4	9	44	7	20	35	11	29	38
80 & above	3	10	30	7	13	54	10	23	43
Total	8	31	26	25	59	42	33	90	37

42% of female mobile phone users). Thirty-seven percent of all mobile phone users in these homes were interviewed (see Table 3).

Findings

Home Environment and the Fixed Line Phone Paging System

One of the contributing factors to why residents of homes for elderly people use mobile phones has to do with the physical setting of the internal environment of the homes. In terms of the number of residents, voluntary sector homes in Hong Kong are quite large, with the number of residents in these eight homes ranging from 126 to 320. They are institutional establishments, like a school or a hospital. As mentioned above, hostels are located on the first or second floor of public housing estates, and care and attention homes are mostly independent purpose-built buildings. There are no fixed telephones in the residents' rooms; instead, such telephones are located either in rooms for common usage or in the staff general office. It could take some time for a resident to answer a phone call because, while a fixed-line phone is usually located on the same floor as the resident's room, it may be necessary for the resident to walk some distance to use the phone. For all in-coming calls, homes have their own practice of paging. Homes have installed a sound system, and there are amplifiers in all or some of the residents' rooms, corridors, living rooms, canteens, activity rooms, and places usually frequented by residents. Whenever there is an incoming call for a resident, the resident's name will be broadcast through the public address system. The resident has to answer the phone outside his/her room. There is also the problem of residents having to queue to use the fixed phone in the home during busy periods. In short, the inconvenience of using the fixed-line phone has prompted the adoption of the mobile phone.

One respondent said that she decided to use the mobile phone after one occasion when she had hurried out of her room to answer the phone and accidentally fell and injured herself. Another said that she did not want people to tell her to answer telephone because it would bother others, and that the mobile phone was far more convenient. Some people do not like to queue up to use the fixed-line phone; others find it difficult to get to the fixed-line phone because of physical frailty, such as reduced mobility and respiratory difficulties.

Comparison of Mobile Phone Users in Hostels and Care and Attention Homes

A comparison of mobile phone users in hostels and care and attention homes shows that the percentage of users in hostels is much higher in every age category. Among hostel residents, 11% are mobile phone users as compared to 4% of residents of care and attention homes. In other words, there is also a tendency among residents with better self-care abilities to have mobile phones (see Table 4).

Sex and Age of Mobile Phone Users

It is assumed that those who own a mobile phone will also use the mobile phone. Based upon this assumption, there are more male mobile phone users than female users, with a result of 10% of resident being male users compared to 5% female users. It has to be noted that out of the eight homes, one is a female home, and there was also one home in which none of the male residents were mobile phone users during the time of interview. In the remaining six homes, there were a higher percentage of male users than female users. It

Table 4. Mobile Phone Owners in Hostels and Care and Attention Homes

Age	Hostel			Care and attention homes			Total		
	No. of mobile phone owners	Total population	(%)	No. of mobile phone owners	Total population	(%)	No. of mobile phone owners	Total population	(%)
60–65	2	5	40	0	6	0	2	11	18
66–70	8	19	42	5	20	25	13	39	33
71–75	16	41	39	7	69	10	23	110	21
76–80	11	84	13	18	223	8	29	307	9
80 & above	9	264	3	14	722	2	23	986	2
Total	46	413	11	44	1040	4	90	1453	6

Table 5. Sex and Age Distribution of Residents Who Use Mobile Phones

Age	Male			Female			Total		
	No. of mobile phone owners	Total population	(%)	No. of mobile phone owners	Total population	(%)	No. of mobile phone owners	Total population	(%)
60–65	1	7	14	1	4	25	2	11	18
66–70	3	19	16	10	20	50	13	39	33
71–75	8	56	14	15	54	28	23	110	21
76–80	9	66	14	20	241	8	29	307	9
80 & above	10	170	6	13	816	2	23	986	2
Total	31	318	10	59	1135	5	90	1453	6

was noticed that more of those under the age of 75 were inclined to use mobile phones. However, among the residents who were interviewed was a 93-year-old ex-farmer, who used a mobile phone.

It can be concluded that users of mobile phones are found among all age groups and sexes, with some concentration in the 66 to 75 age group, and that there are more male than female users (see Table 5).

Previous Occupation of Respondents

Residents who use mobile phones are not a homogeneous group with regard to previous occupation, as indicated in Table 6. This suggests that it is not necessary for older people to have a familiarity with information and communication technologies from their working life in order for them to make

Table 6. Previous Occupation

Occupation	Male	Female	Total	(%)
Housewife	0	9	9	27.2
Factory worker	2	8	10	30.3
Hawker	0	4	4	12.1
Service sector worker	1	3	4	12.1
Clerk	1	1	2	6.0
Restaurant manager	1	0	1	3.0
Self-employed	1	0	1	3.0
Farmer/fisherman	2	0	2	6.0
Total	8	25	33	100

Table 7. Length of Mobile Phone Usage

	Male	Female	Total	(%)
Less than a year	2	6	8	24
1–2 years	1	11	12	36
3–4 years	4	7	11	33
5–6 years	1	1	2	6
Total	8	25	33	100

use of mobile phones. This is contrary to what Oksman found in her study in Finland (Oksman 2004).

Length of Mobile Phone Usage

Among the eight male respondents, only two were using a mobile phone for the first time. As for the rest, the average number of years of mobile phone usage was three years. Among the 25 female mobile phone users, eight had used a mobile phone for less than a year; eleven had used a mobile phone for about a year, six for 2 years, four for 3 years, one for 4 years, and one for 5 years. There were slightly more female mobile phone users in the last 1 to 2 years. We can see from these figures that 24% of respondents had a mobile phone for less than 1 year, 48% for 2 years, 36% for 1–2 years, 33% for 3–4 years, and 6% for 5–6 years (see Table 7).

Who Bought the Mobile Phone?

Two-thirds of the mobile phones had been purchased by the children of the residents and two (6%) had been a gift from the residents' grandchildren (see Table 8). One woman, whose husband had died and who had no children, had

Table 8. Who Bought the Mobile Phone?

Son or daughter	21	63%
Son-in-law	2	6%
Grandchildren	3	9%
Old phone from niece	1	3%
Old phone from godson	1	3%
Old phone from ex-colleague	1	3%
Self	4	12%
Total	33	100%

Table 9. Who Pays the Network Service Fees?

Son or daughter	22	66%
Son-in-law	2	6%
Daughter-in-law	1	3%
Grandchildren	1	3%
Niece	1	3%
Godson	1	3%
Self	5	15%
Total	33	100%

been given her phone by her niece. One resident had been given his phone by an ex-colleague who sometimes had difficulties in finding part-time staff.

Who Pays the Network Service Fees?

Only five residents (15%) paid the network service fees themselves; with family members paying the fees for the rest (see Table 9). The monthly fees ranged from a minimum of HK$40 to a maximum of HK$150. Some respondents used prepaid cards instead of paying a monthly fee; these can cost as little as HK$100 and last for several months.

Frequency of Communication with Family

Residents used their mobile phones quite frequently: 39% phoned everyday, 15% phoned 3–6 times a week, 21% phoned 1–2 times a week, while 24% called occasionally (see Table 10). Among the respondents, there was the "daily contact" type, "hello and goodbye" type and "frequent talker" type. There was also the type of user who treated the mobile phone as a device for making emergency calls only. Ten of the 13 "daily contact" groups were female and the remaining three were male. Women were slightly more likely than men to be daily users.

Table 10. Frequency of Communication with Family

Frequency	Male	(%)	Female	(%)	Number of users	(%)
Daily	3	37.5	10	40	13	39
3–6 times per week	1	12.5	4	16	5	15
1–2 times per week	1	12.5	6	24	7	21
Occasionally, not regularly	3	37.5	5	20	8	24
Total	8		25		33	100

The Contents of Conversations: More Than Reporting Safety

There is an assumption that mobile phone usage among older people has more to do with reporting their safety than with socializing, as is the case with teens (*New York Times 2004*). The present study found that mobile phone usage reflects certain family and social relations and that the contents of conversations are about more than reporting safety. The contents of mobile phone conversations are diverse, but mainly concern their families, health issues, news about the home, and other activities.

The research identified five kinds of "talk." First of all, there is the specific kind, for example, where a daughter asks her mother to return home to take care of the grandchildren or to help out with household chores when she goes out to work. One resident had to phone another home in which her husband lived every day to make arrangements for taking care of him. It is interesting to see a continuation of the family roles that residents performed prior to their admission to the home. Second, there are "ordinary family chats," which are mainly concerned with finding out how the residents are getting on, how their grandchildren are doing, how they are keeping, and exchanging words of comfort. The third type of "talk" is for making arrangements to visit the home, to meet for lunch or other meal appointments, and making doctor's appointments. Fourth, is the kind that is about informing the family of news in the home for older people. Fifth, 21 out of the 33 interviewees said that they would use their mobile phones for purposes other than to contact their family, including talking to brothers and sisters, relatives, friends, for entertainment, and business. One respondent said that she used her mobile phone for the "Mark Six" (lottery) and one respondent said that she use it to arrange her volunteer work. Apart from local calls, the children of five respondents had emigrated aboard, and they had to call their children long distance; this was far more convenient than using a fixed phone. The sons of two residents work in China, and the mobile phone helped the residents to stay in touch with them.

Table 11. Talking to Family Members is:

Necessary	23	70%
Good	15	45%
Helps me to feel better	11	33%
Leads to more quarrels	1	3%

Note: Respondents were able to choose more than one answer

The Importance of the Use of Mobile Phones to the Respondents

All of the respondents were positive about the use of mobile phones, as indicated in Table 11. Even one respondent who said she had "more quarrels," agreed that talking to family members is also "necessary," "good," and helped her to "feel better." One respondent said the time she spent talking to her family was precious.

Mobile Phone Design: User-friendliness

Regarding the design of mobile phone and the aspect of user-friendliness, 7 out of 33 of the respondents said that they initially resisted using a mobile phone, the reasons being that the phone was too heavy, too expensive, or that the condition of their eyes or ears was both fair.

Most of the respondents had no problem learning how to use a mobile phone. Only 2 out of the 33 respondents had difficulties, and 2 of them only know how to receive calls but not to make calls. When asked how to overcome the difficulties of using the phone, they said they would ask staff at the home or security people to help. People from the management of the homes were asked whether they had observed any problems with the residents using mobile phones. The answers were very positive, with all of the homes responding that there were no problems, no cases of stolen phones, no losses within the home, and only a few cases of loss when residents took their phones outside.

Difficulties regarding the use of mobile phones mostly involved motor coordination of the fingers, and visual and auditory abilities. Some said that they found it difficult to read messages, to switch on the vibrating alert, or to store information. Usually, all of the residents can have their mobile phones re-charged near their beds. But some homes charge residents a monthly fee for using extra electricity.

Many brands of mobile phones were used among the respondents; the users seemed to have no favorite brand or model. After all, the mobile phones had mostly been bought by the users' families. There was no favorite network provider, with cost being the main concern.

We asked the elderly respondents how the designs of mobile phones can be improved. The responses include having bigger words, bigger screens, louder sound, lighter weight, buttons that are easier to press, and making the phones easier to operate. Some said that there should be a model just for the elderly.

In sum, apart from considering the above suggestions, the new mobile phone for elderly people can also work in combination with other types of technology (Ostlund 2004) such as the personal emergency link, a device that can be used to track and locate the position of elderly people if they have problems such as dementia.

Conclusion

An interesting aspect of the mobile phone is that, like any other kind of technology, it both empowers and disempowers. Questions relating to domestication and routinization (Ling 2004) are important for users. For this preliminary study, the following themes are worth examining.

A mobile phone is a practical material device but communication is abstract and also symbolic and interactive. Receiving a used mobile phone from one's close kin represented care and concern. The sentimental aspect of the mobile phone as a gift was quite obvious in this study.

It is important that older people can be placed in homes in their own communities so that even though they are living physically apart from friends and family members, they can still visit one another. The mobile phone allows visiting arrangements to be made in a far more efficient manner than would otherwise be possible. The study revealed the examples of a 76-year-old ex-fisherman who was able to maintain his links with his previous community and visit his friends, and a 93-year old ex-farmer who was able to meet his family in a tea house several times a week through the use of the mobile phone.

At present there is little provision for couples to stay together in the same home, and even if they do, they are usually accommodated separately. This policy, however, needs to be reconsidered. One resident in this study had to travel daily to take care of her sick husband in another home, which was not only a waste of her time but also an additional hardship. This finding was not expected, but it certainly reflects the importance of family relations and the use of the mobile phone. Wives and husbands should be allowed to stay in the same home if possible so that they can take care of each other (Race 1982; Tsang 2001). The current policy is unnecessarily exclusive.

In general, nowadays, mobile phones are used not only by the privileged. Mobile phones and network service have become so affordable that mobile phone use is rising. We are also seeing families who have given their old phones to elderly members of their family when they switch to newer models. The cost of overseas calls has become increasingly affordable. Those who have family members living overseas need a mobile phone to be able to maintain direct contact with them. The physical environment of old people's homes, the paging system of the fixed-line phone, coupled with certain physical and health limitations all contribute to an increase in the use of the mobile phone.

Although this study is exploratory in nature, it identifies key characteristics of mobile phone users who are under residential care in Hong Kong. Older mobile phone users in homes for the elderly are still able to maintain their own networks outside their homes. Some of them are highly active, and most of them are regular mobile phone users; only a small handful of them treat the mobile phone merely as a device for making emergency calls (Ling 2004). It was also noticed that in addition to a means of communication that is universal, the mobile phone can also represent care and concern by the families of the elderly residents.

The changing structure of Hong Kong society has had an impact on the family structure. Families are shrinking in size, and if support networks are available, they tend to be limited (Ngan 1990). The mobile phone as a device is able to create a "cyber space" between the generations. Through mobile phone communication, a familial space is created between parents and their children or grandchildren, even though they may not be living together. Hearing familiar voices caused worries to be dispelled, and in such a way that both parties can carry on. This enables elderly people as well as their children to preserve their independence and sense of respect.

The mobile phone promotes family resilience. Care may not be that abundant but it is available, and it is not face-to-face talking but voice-to-voice words of care that count.

References

Asia Business Consulting. 2004. http://asiabusinessconsulting.blogspot. com/2004_08_01_asiabusinessconsulting_archive.html.

Census and Statistics Department. 2001. *Hong Kong Population 2001 Census.* Hong Kong: Hong Kong Government Printer.

Chan, A. C. M. and D. R. Phillips. 2002. "Long Term Care Policy in Hong Kong." In *Ageing and Long-Term Care: National Policies in the Asia-Pacific*, edited by

D. R. Phillips and A. C. M. Chan. Singapore: Institute of Southeast Asian Studies & International Development Research Centre.

Chow, N. W. S. 2000. "Ageing in Hong Kong." In *Ageing in the Asia-Pacific Region: Issues, Policies and Future Trends*, edited by D. R. Phillips. London: Routledge.

Hong Kong Government. 1977. *Green Paper on Services for the Elderly.* Hong Kong: Hong Kong Government Printer.

Hong Kong Government. 1979. *White Paper on Social Welfare into the 1980's.* Hong Kong: Hong Kong Government Printer.

Kwok, H. K. 2001. "Family Support Network and Community Development in Hong Kong." In *Aging, Gender and Family in Singapore, Hong Kong and China*, edited by K. W. K. Law. Taiwan: Academia Sinica.

Lee, R. P. L., J. J. Lee, E. S. H. Yu, S. G. Sun and W. T. Liu. 2000. "Living Arrangement and Elderly Care: The Case of Hong Kong." In *Who Should Care for the Elderly? An East-West Value Divide*, edited by W. T. Liu and H. Kendig. Singapore: Singapore University Press.

Ling, R. 2004. *The Mobile Connection: The Mobile Phone's Impact on Society.* San Francisco: Morgan Kaufmann Publications.

Ming Pao. 2004. "Old People Giving up Using Fixed Line Phone." March 1st. (Hong Kong Chinese newspaper).

New York Times. 2004. http://www.nytimes.com/2004/10/11mobile. hhtml?th

Ngan, R. 1990. "The Availability of Informal Support Networks to the Chinese Elderly in Hong Kong and Its Implications for Practice." *Hong Kong Journal of Gerontology* 4:19–27.

Oksman, V. 2004. "Mobile Communication in the Everyday Lives of Finnish Senior Citizens: The Aspects of Safety, Lifestyle and Health." Paper Presented at the Workshop on Mobile Technologies and Health: Positive and Negative Aspects, June 7th–8th, 2004, Udine, Italy.

Ostlund, B. 2004. "Mobile Telephony and Active Ageing — Swedish Examples of Policy and Practice" Paper Presented at the Workshop on Mobile Technologies and Health: Positive and Negative Aspects, June 7th–8th, 2004, Udine, Italy.

Race, D. 1982. "Residential and Institutional Services for the Elderly." Resource Paper No. 3, Department of Social Work, University of Hong Kong.

Social Welfare Department. 2005. Hong Kong: Hong Kong Government. Information about elderly services in Hong Kong. http:www.info.gov.hk/swd/text_eng/ser_sec/ser_elder/index.html

Tsang, F. Y. 2002. "Elderly Services in Hong Kong: From the Choice and Autonomy Angle." In *Advances in Social Welfare in Hong Kong*, edited by D. T. L. Shek, L. M. Chow, C. F. Au and J. J. Lee. Hong Kong: The Chinese University Press.

Working Party on the Future Needs of the Elderly. 1973. *Services for the Elderly.* Hong Kong: Hong Kong Government Printer.

Working Group on Care for the Elderly. 1994. *Report of the Working Group on Care for the Elderly.* Hong Kong: Hong Kong Government Printer.

Part 3
ICTs and Work

9 The Impact of Internet Use on Transnational Entrepreneurship: The Case of Chinese Immigrants to Canada

Wenhong Chen

Introduction

Immigrants are a large, growing, and dynamic source of entrepreneurship (Portes 1995; Menzis *et al.* 2002; Li 2002). Moreover, because they maintain ties abroad they are likely to take advantage of modern transportation and communication technologies and to extend their businesses transnationally. Transnationalism has significant consequences for immigrants, their communities, and both the host and home country (Levitt *et al.* 2003). In the host country, transnational entrepreneurship offers immigrants an alterative approach to economic incorporation (Guarnizo *et al.* 2003). In the home country, the transitional economic activities of immigrants have a profound impact on social and economic development.

Research on immigration transnationalism has been growing rapidly since the 1990s. In response to critiques that transnationalism is not a new phenomenon, scholars refer to the use of new technologies as one fundamental feature of contemporary transnational practices. The existing literature stresses that the convenience and simultaneity afforded by modern technologies significantly intensify and expand transnational activities (Portes *et al.* 2002). Furthermore, the Internet and other ICTs open up cyberspace for transnationalism because online communication and exchanges can crisscross national boundaries "without actual bodily movement" (Smith and Guarnizo 1998:14). While the impact of modern transportation and communication technologies on transnationalism is frequently mentioned, there has been little systematic research on how immigrants actually use new technologies to participate in transnational activities.

On the other hand, literature on the social impact of the Internet has addressed substantial topics such as the patterns, causes, and consequences of the digital divide within and across countries, the interaction of computer networks and social networks, and the dynamic relationship between online and offline activities. These insights are valuable when we look at the impact of the Internet on transnational entrepreneurship, even though few studies in this line of literature have focused on immigrant groups.

At the intersection of these two literatures — immigration transnationalism and the interaction of technology and society, the impact of Internet use on transnational entrepreneurship provides a unique vantage point from which to observe the ways in which social, cultural, and technological forces affect economic actions. Drawing on data collected through participant observations and in-depth interviews in Toronto and Beijing, this chapter explores the socio-technological dimension of transnational entrepreneurship. The chapter is organized as follows. The first section provides a brief review of the literature on immigrant transnationalism and the literature of the social consequences of the Internet. In particular, the review focus on how these two literatures can inform each other to improve our understanding of the impact of Internet use on transnational entrepreneurship. The second part presents data collected from Chinese immigrant entrepreneurs living in Toronto and Beijing. The results indicate that Internet use has a profound impact on transnational entrepreneurship in terms of searching for information, networking, and mobilizing resources. However, the Internet is not everything, as transnational entrepreneurs selectively use multiple forms of communications media, in accordance with whichever is most convenient and socially appropriate.

The Blind Spot — Immigrant Transnational Entrepreneurship and Internet Use

Technology Use in the Literature on Immigrant Transnationalism

The term "transnationalism" has gained currency since the early 1990s (for the latest reviews, see Portes 2003; Waldinger and Fitzgerald 2004). Transnationalism is theorized as a response to the logic of global capitalism and flexible accumulation, and a grassroots resistance to the transnationalism from above controlled by states and multinational corporations (Levitt, 2001; Smith and

Guarnizo 1998). Transnationalism is conceptualized as the process "by which immigrants forge and sustain multi-stranded social relations that link together their societies of origin and settlement" (Basch *et al.* 1994:7). It presents a new perspective with which to study the enduring connections and participation of immigrants in the economic, political, cultural, and religious activities of the sending countries. Transnational entrepreneurs are defined as "self-employed immigrants whose business activities require frequent travel abroad and who depend for the success of their firms on their contacts and associates in another country, primarily their country of origin" (Portes *et al.* 2002:287). Research on Latino immigrants in the United States has shown that a large proportion of immigrant entrepreneurs[1] engage in transnational business.

Critics claim that the literature on immigrant transnationalism tends to overemphasize the novelty, magnitude, and sustainability of transnationalism. They argue that transnational activities have existed throughout history; that only a small proportion of immigrants engage in transnational activities; and that transnationalism will decline or even disappear in the second generation. More recently, Waldinger and Fitzgerald (2004) question the existing literature for downplaying the role of the state and multiple political constraints in constraining transnational activities. In response, it is conceded in the more recent literature on immigrant transnationalism that the phenomenon may not be new. Yet, transnational activities documented in history neither had the kind of regularity nor reached the critical mass that characterizes contemporary transnationalism. More importantly, the diffusion and use of new technologies is a necessary condition and a fundamental feature of contemporary transnationalism (Portes *et al.* 1999; Portes *et al.* 2002; Levitt 2001; Vertovec 2003; Gold 2001). Accordingly, transnational practices cannot proliferate without advancements in technology. The following quotation is a case in point:

> It was not possible for would-be transnational entrepreneurs to travel to Poland or Italy over the weekend and be back at their jobs in New York by Monday. Nor would it have been possible for leaders of an immigrant civic committee to keep in daily contact with the mayor of a Russian or Austrian town in order to learn how a public works project, financed with immigrant money, was progressing. Communications were slow and, thus,

[1]Close to 60% of self-employed immigrants surveyed by the Comparative Immigrant Entrepreneurship Project (CIEP) have expanded their businesses transnationally. However, this comprises about 5% of all immigrants (Portes *et al.* 2002).

> many of the transnational enterprises described in today's literature could not have developed (Portes *et al.* 1999:223).

The existing literature stresses that the convenience and simultaneity afforded by modern transportation and communication technologies expand and intensify transnational activities to an extent that is incommensurate with transnational activities in history (Portes *et al.* 2002). Modern technologies make contemporary transnationalism possible by facilitating the building of transnational communities and the maintenance of membership in such communities (Smith 1998). For instance, Levitt argued that new technologies "transform migrants' contact with their sending communities, allowing them to be actively involved in everyday life in fundamentally different ways than they could have been in the past" (2001:203).

Furthermore, the existing literature asserts that the Internet and other ICTs open up cyberspace for immigrant transnationalism because online communication and exchanges can crisscross national boundaries "without actual bodily movement" (Smith and Guarnizo 1998:14). Portes and colleagues further extrapolated that "obviously, the space-compressing power of modern electronics allows persons who have command of these resources to engage in transnational activities without the need for face-to-face contact. Hence, the barrier of distance gradually diminishes as communities become able to substitute traditional personal contact with new electronic means of communication" (1999:224).

Literature on the Social Impact of Internet Use

In the early years of the Internet, pundits either announced the advent of a brave new world of cyberspace or warned about the alienation of humankind through the smart machine. The social impact of Internet use was the subject of speculation and anecdote rather than of systematic analysis and examination. However, starting in the late 1990s and especially after the dot.com mania in 2000, there has been systematic documentation of who are surfing the Internet and what online activities they are involved in. Scholars have come to adopt a more realistic stance in examining how the Internet affects individuals, communities, and societies. Empirical studies have taken the place of anecdote and speculation. These studies have addressed substantial topics such as the patterns, causes, and consequences of the digital divide within and across countries, the interaction of computer networks and social networks, and the dynamic relation between online and offline activities. These insights are valuable when we look at the impact of the Internet on transnational entrepreneurship, even though little of the research in this line of literature has focused on immigrant groups.

First, many studies rest on the belief that the informed use of the Internet is already widely available and growing swiftly. Portes and colleagues assumed that "the ready availability of air transport, long-distance telephone, facsimile communication, and electronic mail provides the technological basis for the emergence of transnationalism on a mass scale" (1999:223–224). By assuming that immigrants are able to access and use the Internet for transnational purposes, there is a tendency to ignore the issue of the digital divide — the gap splitting individuals and societies that have the resources to participate in the information era and those that do not. The digital divide is particularly deep and pervasive among ethnic minorities/immigrants and the countries from which they come. Not all immigrants have access to new technologies. For instance, according to a report by the World Economic Forum (2002), Finland has more Internet hosts than all of the Latin American countries put together. While close to 60% of Americans had access to the Internet, less than 5% of Mexicans went online in 2002. Within national states, the uneven diffusion of the Internet appears along the familiar lines of social inequalities such as socioeconomic status, gender, and ethnicity (Chen and Wellman 2005).

As important, capitalizing on the Internet depends on who uses it, for what purposes, and under what circumstances. The digital divide persists even after former information have-nots have gained access to the Internet because computer and cognitive skills are required if the Internet is to be used meaningfully and productively, such as to seek information, develop community networks, construct social capital, and participate in political activities (Hargittai 2002). Socioeconomic status (such as education and occupation), gender, and ethnicity are related to the extent to which people can benefit from Internet use (DiMaggio *et al.* 2001; Haythornthwaite and Wellman 2002; Haythornthwaite and Hagar 2004). For instance, until recently, the contents of the Internet were more likely to target well-off, well-educated, and English-speaking users. Immigrants and minorities are more likely to have difficulties obtaining relevant practical and local information on the web in their own languages, such as job listings, housing listings, and local events.

Second, there is a lack of studies on how new technologies affect the networks of immigrants in the process of transnationalism. The proliferation of the Internet and other ICTs affords new practices of social interaction. Various studies have found that Internet use is positively related with social integration and community building (Howard 2004; Wellman 2002; Quan-Haase *et al.* 2002). Sociable and meaningful relations can be maintained through the

Internet and serve as a source of social support (LaRose *et al.* 2001). Reciprocity and attachment to virtual communities could unfold over time (Miyata 2002). Consequently, Internet use has a positive impact on the accumulation of social capital (Katz and Rice 2002; Wellman 2001). Over the years, various formulations of transnationalism have been developed. Border-crossing networks have been viewed as one fundamental characteristic of immigrant transnationalism (Glick-Schiller *et al.* 1992). Accordingly, networks are considered to be among the best sites to study transnational practices (Vertovec 2003). Yet the interaction of computer-mediated communication and social networks has yet to enter the discussion. We know little about how Internet use affects social networks in the process of transnationalism.

As the Internet has become an integral part of everyday life, it is not appropriate to treat online and offline communication as two separate, mutually exclusive worlds (Haythornthwaite and Wellman 2002). Yet, in asserting that online communication may replace in-person contact, the existing literature on transnationalism tends to overestimate the impact of the Internet. While the Internet enables the geographic boundaries of social networks to expand, this does not mean that local interaction has lost its significance — individuals also make connections within their local setting defined by geographic, organizational, or ethnic boundaries (Calhoun 1998). Far from being substituted by online communication, in-person experiences continue to be valued. For instance, most strategic business relationships are local because of the importance of timeliness and the necessity of face-to-face contact (Saxenian 1999, 2002). Business people and professionals often travel long distances to hold frequent in-person get-togethers. As a result, online relationships are articulated and intertwined with offline relationships. Furthermore, there are other information and communication technologies, each with different social affordances. The Internet is embedded in the broader context of the use of media and technologies. Media multiplexity captures the phenomenon of people selectively using forms of multiple communications media in accordance with whichever is most convenient and socially appropriate — email, telephone, and face-to-face interaction — to keep in contact with family, friends, and workmates (Haythornthwaite 2001). Cyberspace is not an isolated place segregated from the physical place.

The literature on transnationalism has highlighted the social affordances of modern transportation and communication technologies that intensify and expand transnational activities. New communication technologies give immigrants access to images and sounds from their places of origin or from the global diasporic community, enabling them to simultaneously take part in the social life of both the host country and the home country (Georgiou 2003). Immigrants use the Internet (online newsgroups, list-serves, mailing lists, etc.) to construct public forums where they can socialize and discuss issues. The Internet also facilitates the emergence of new cultural spaces and thus fosters stronger and richer connections with the homeland and diasporic communities elsewhere (Miller and Slater 2000; Panagakos 2003; Mitra 2003; Yang 2003). However, there has been little empirical research on how new communication technologies are transnationlizing immigrant life. Too often, new communication technologies are conceptualized as backdrops. Even fewer studies have explored how immigrants use new technologies in entrepreneurial activities. This chapter attempts to look into this blind spot and address the question of how the Internet affords transnational entrepreneurship.

Research Method

The data for this paper were collected through in-depth interviews and participant observations in Toronto and Beijing from 2002 to 2004. By the end of 2004, I had conducted 16 formal in-depth interviews and 32 informal interviews in Toronto; and 32 formal in-depth interviews and 14 informal ones in Beijing. Using a purposive sampling technique, I selected interviewees who were transnational entrepreneurs or knowledgeable about transnational entrepreneurship (including activists in business or professional associations, journalists, venture capitalists, bankers, and governmental officials). The interviewees were recruited through various sources — professional and business associations, governmental business promotion agencies (both Chinese and Canadian), the Chinese language media, business events and networking activities, business directories posted on the websites of industrial incubators in China, and snowball referrals. In both cities, one central theme of the in-depth interview was how actors in the transnational field use new communication technology to engage in entrepreneurship. All of the formal interviews were guided by an interview outline. The interviews ranged from 1 hour to 4 hours, and were conducted either in the interviewee's office or at a public place such as a café or teahouse. Interviewees could choose to be interviewed in English or Chinese. However, with the exception of three interviews that were conducted in English, all were conducted in Mandarin. All of the interviews were transcribed.

In the two cities, I carried out 34 participant observations at a variety of business and community events and activities, particularly those related to entrepreneurship. All field observations were documented through written field notes. In Toronto, these events included business forums and seminars organized by various immigrant business or community associations (within or beyond the Chinese community), activities organized by firms owned by Chinese immigrants to Canada, and activities organized by Canadian governments at the federal, provincial, and city levels. In Beijing, these events ranged from trade shows, business promotions, and exchanges at various industrial incubators affiliated with the Zhongguancun Science Park in Beijing, returnee professional associations, and Canadian business or community associations. In addition, the Internet has been an important research site for this study. I have been monitoring major Chinese language websites and newspapers in Toronto since 2002.

In the pages that follow, I will first discuss the social context that induces immigrant entrepreneurs to engage in transnational activities. I then move on to examine how the Internet affects the networks of immigrant entrepreneurs and facilitate their involvement in transnational entrepreneurship.

The Rise of Transnational Entrepreneurship

The Growth of the Immigrant Population

Although new technology can induce transformations in production/consumption, power, experience, and culture (Castells 1996), the extent to which a technology lives up its potential is contingent on social and institutional contexts (Agre 2002). In a similar vein, the impact of the Internet on transnational entrepreneurship is embedded in social and institutional contexts. The macro conditions set the stage for transnational practices. Without the Chinese government's relaxation of restrictions on the emigration of the country's citizens, its development strategy encouraging foreign investment, or the Canadian government's liberal immigration policy, there would have been little room even for the most creative and networked of entrepreneurs to engage in transnational activities between the two countries.

In 1967, Canada adopted the "points" system to select business and skilled immigrants. The country welcomes immigrants to offset its aging population and a deficit in skilled labor. Yet it was not until the late 1970s when the PRC (People's Republic of China) government began to launch reforms that the restrictions imposed on the emigration of the country's citizens were gradually relaxed. The number of PRC immigrants to Canada has increased dramatically since the 1990s. While Hong Kong was the primary source of Chinese immigration from the end of World War II to the mid-1990s, the PRC has taken over to become the largest source of immigration to Canada since 1998. From 1996 to 2002, about 200,000 immigrants from the PRC landed in Canada. More than two-thirds of them are in the skilled worker class, accounting for 17% of all skilled immigrants to Canada in the same period. There has also been a remarkable increase in business immigrants from the PRC, who made up close to 30% of all business immigrants to Canada in 2001 and one quarter in 2002.

PRC immigrants are concentrated in major Canadian metropolitan cities. Toronto is the most powerful magnet for new immigrants from China. On average, about half of all PRC immigrants who landed in Canada between 1996–2002 selected Toronto as their new home. The increase in the immigrant population and its urban concentration have created favorable conditions for the growth of ethnic economies, such as protected markets, co-ethnic labor, customers, media, banking, and business associations (Fong and Ooka 2002; Zhou and Tseng 2001).

The policies of the state have significantly affected the immigrant population in Canada. The state, especially the state of the sending country, also plays a major role in promoting immigrant transnationalism. The Chinese government has been keen to attract immigrant entrepreneurs to return and do business. Furthermore, the state and its functionaries are deliberately using new technologies to project an image of modernity and efficiency. Professional and business associations in the immigrant community are also active in supporting the transnational entrepreneurship activities of their members.

The Growth of the Ethnic Economy

Although many new immigrants are university educated and proficient in English, they are more likely than native-born Canadians to be unemployed or underemployed. Their relative income to native-born Canadians has continued to decline since the 1980s (Reitz 2001). The prolonged process of immigrant integration is a huge loss to the Canadian economy, a social cost to the welfare system and social services, and a personal cost to immigrants and their families. Entrepreneurship seems to provide immigrants with an alternative way to participate in the labor market.

In fact, Chinese immigrant entrepreneurship has been rising in Canada since the last two decades (Li 1993). Chinese immigrants have been found to have a higher rate of self-employment than other immigrant groups. The proportion of self-employed among all foreign-born Chinese Canadians rose from 3.6% in 1981 to 7.4% in 1996 (Li 2002). Many immigrant entrepreneurs are involved with transnational economic activities. For instance, Hiebert and

Ley (2003) reported that 13% of the 164 Mandarin-speaking immigrants they studied in Vancouver have a business in their home country, 28% have jobs that require them to travel frequently to their home country, and 42% have property in their home country.

By definition, entrepreneurs are brokers. They "seize opportunities for purposive intervention by maneuvering back and forth between different social networks as well as cultural or social-psychological settings" (Emirbayer and Mische 1998:1007). Simultaneously living in two societies, the social, cultural, and geographical distances between North America and China create opportunities for immigrant entrepreneurs to identify market niches, to manipulate structural holes, and to act as brokers. The type of transnational business (such as transnational media productions, education, immigration services, telecommunications, or tourism) suggests that many such businesses are based on the trading of information, goods, services, technologies, knowledge, and experiences between the country of origin, the country of destination, and/or a third country. For example, together with her husband and another three IT engineers with extensive North American experience, DLL co-founded a high-tech firm with offices in Beijing and Toronto. She reflected on how they ended up with an IT training business:

> Talking with friends and meeting firms in the industry, I found that there is a great demand for good programmers [in China]. Many software firms are doing contract work outsourced from North America. However, there are few instructors with hands-on North American programming experience. Isn't that a good niche for us? ... So, instead of entering the competitive market of software programming that requires more investment, we reinvented ourselves as an IT training company. (Beijing, 2004).

The Internet and Chinese Language Online Media

Academic research on Internet use among immigrants is hard to find. However, many facts suggest that Chinese immigrants are intensive users of modern technologies.[2] Asian-Americans have been the heaviest users in the U.S., with the longest Internet experience. They are more likely than other ethnic groups to use the Internet for work or school-related purposes (Spooner 2001). In

[2]Besides technological advancements, scholars have also pointed out other ways in which the transnationalism of today is qualitatively different from the kinds documented in the past, including the shift from assimilation to cultural pluralism in the receiving societies; the formation of national identity in the sending countries; and the emergence of a new international human rights regime (Waldinger and Fitzgerald 2004:1187).

Toronto, 86% of Chinese households owned a computer in 1999. Close to 80% of them had Internet access (ACNielsen 1999). Immigrant entrepreneurs are taking advantage of the surge in Chinese Internet users, as exemplified in JL's case.

Migrating to Toronto in the mid-1990s after six years of Ph.D. studies in Scotland, JL found a job as a programmer in a local telecommunications firm. While the pay was not bad, JL was inspired to become an entrepreneur when volunteering in a Chinese business association. There were many Chinese language Internet portals and websites both in China and in the global Chinese diasporic community by the end of the 1990s. However, in Toronto, there was only one Chinese language Internet portal serving immigrants from mainland China. Teaming up with six friends, he decided to build an "online community" for the mainland Chinese community in Toronto. Following the bursting of the dot.com bubble, many thought they were doomed:

> A good friend advised me against the idea. He showed me a website called "dot.com graveyard." The website has a gravestone with the name and life of each dead dot.com and the last word of the founder. I was not disheartened. Skilled immigrants from China are coming and they use the Internet a lot. The first thing they do after finding a shelter is to buy a computer and get Internet access. As long as we provide content relevant to the community, our website can be sustained and grow (2004, Toronto).

When I interviewed JL in December 2004, his website had one of the largest readerships among Chinese language websites in Toronto. JL himself has become a "network hub" in the Toronto Chinese community, as the website offers him a platform to get to know and to link up with different groups of people — leaders of business associations and immigrant service agencies, professionals, entrepreneurs in Toronto, as well as visiting officials and business people from China.

Websites run by and for Chinese immigrants have been booming in North America. There are at least 10 websites run by and for immigrants from the PRC in Toronto, such as Toronto Chinese Online (http://www.torcn.com), Carefree in Canada (http://www.51.ca), It 4 U (http://www.it4u.net), Toronto Information Hub (http://www.torontoservice.com), and so on. In addition, almost all major Chinese-language newspapers have gone online. While Chinese-language websites in China, Hong Kong, Taiwan, or elsewhere are just a click away, North America-based Chinese language media not only keep immigrants well-informed of their home country but also provide them with a roadmap to integrate into the host society (Zhou and Cai 2002).

These media outlets themselves are transnational businesses because an important part of their content is from homeland media outlets. Furthermore, advertisements of Chinese ethnic businesses are their major source of income. More importantly, these online media facilitate transnational entrepreneurship in important ways. First, entrepreneurship and local business events are an important part of the content of Chinese-language media, both online and offline. For example, the *World Journal*, one of the big three Chinese dailies in Canada, has a column on Fridays on entrepreneurship written by a renowned accountant and former president of a Toronto Chinese business association. Profiles of successful immigrant entrepreneurs, especially those who have made it either in the mainstream society or back in the countries of origin, regularly appear in various newspapers and websites, under rubrics such as "the Chinese Light" or "the Star of the Community."

Second, most Chinese language websites maintain an entrepreneurship archive. For example, Toronto Chinese Online (http://www.torcn.com/node.php?id=2041) has an "Online Entrepreneurship Bible." Carefree in Canada keeps an "Online Entrepreneurship Dictionary" (http://www.51.ca/info/classdisplay.php?s=&classid=15). These "bibles" or "dictionaries" contain a variety of information and advice on entrepreneurship, such as how to establish a business, small business taxation, custom regulations on import sand exports, business etiquette, and so on.

Third, there are online forums dedicated to entrepreneurship. China Smile is one of them (http://www.chinasmile.net/forums/forumdisplay.php?f=4). From February to May 2004 alone, there were 584 themes and more than 4,000 posts in this online entrepreneurship forum. The topics ranged from how to register a business in Toronto, how to conduct import and export activities between Canada and the U.S., how to enter the mainstream market, to differences between Chinese and North American business cultures. Online entrepreneurial forums, chat rooms, and list-serves can help immigrants build up transnational networks by developing ties in the new country, as communities of practice emerge on list-serves and online forums. The following quote from an interview with KZ is a case in point:

> Many small businesses in my industry use Cisco hardware and software. Therefore, there is a Chinese language online forum of Cisco equipment users. Once I sent an email to the list, asking if anyone knew how to solve a technical problem. I got several responses, saying, 'I don't have the answer to your question. Yet, based on the complexity of your question, you seem to know the stuff pretty well. So I'd like to ask you a question'. As the whole forum saw my answers to these questions, I became, much to my own

> surprise, one of the experts, and people keep sending me questions off the list (2004, Toronto).

The Impact of the Internet on Transnational Entrepreneurship

Information Searching

A border-crossing network is important to transnational entrepreneurship. The construction of such a network involves a deliberate discourse of culture and identity (Ong and Nonini 1997). To build border-crossing social networks, entrepreneurs need a working knowledge of the societies and cultures within which they operate. However, having lived in the home country and the host country does not necessarily mean that most immigrants are "adept in both cultures with the capacity to interpret them to each other and to bring them together" (Lever-Tracy and Ip 1996:247). In fact, many immigrants are encapsulated in their ethnic community, voluntarily or not. They often do not have enough knowledge to function in mainstream society. On the other hand, immigrants who have lived in the host country for a long time may lose a sense of what is going on in the home country, especially in the case of a transitional society like China, which has been undergoing rapid social and cultural changes in the last three decades. Emailing and seeking information are the two most frequent Internet activities in many countries. The Internet helps immigrant entrepreneurs catch up and keep their cultural capital up to date, as shown in the following case.

YF had spent most of his years in North America in "mainstream circles." In his own words, "I got an MA from a good university in the U.S., migrated to Toronto, worked in a big mainstream company, and hung around primarily with the locals..." After the birth of his second child and the immigration of his parents from China to join him in Toronto, YF felt the urge to start up his own business to "feed all these people." However, in the year 2000, when he went to China after a long absence to launch his logistics management software in the Chinese market, he was embarrassed to discover that his knowledge of the country was outdated.

> That was three years ago. I came back for the first time in 10 years. I brought a project based on SAP[3] with me. I thought people would be very impressed. Instead, they laughed at me, saying 'Come on, we have SAP. We know how to use it. Do you have anything more exciting?' I drew the lesson that you really have to know what is needed most in China. In order to catch up, [after

---

[3]SAP is the largest available enterprise application software.

> I returned to Toronto], I began to surf Chinese-language websites every day. I read all kinds of news — social, political, business, science and technology, even entertainment (2003, Toronto).

Another example is from QG, whose Beijing-based firm focuses on developing environmental technologies. At the same time, QG is working on a part-time Ph.D. at the University of Calgary. He said, "The R&D I am doing here will be part of my thesis." An avid Internet user, he described how searching for information on the web is instrumental for the operation of his company:

> From the technical aspect, I use the Internet to search references. In terms of archiving references, China is still behind Western countries. But as a student, you log on the university's e-resource website, do a search, and you get many references. It is not this easy in China yet. I got almost all the information I needed from the web to find out if our ideas had been already tried out by competitors. Only when it is absolute necessary will I visit the Patent Bureau to investigate the details. From the marketing aspect, I search for information about pricing, products, and services provided by our competitors. I also search for client information on the web before I approach clients (2004, Beijing).

Building and Maintaining Transnational Networks

Immigrants rely on connections in their home and host countries to expand their businesses transnationally. Affording easy, rapid, and border-spanning communications, the Internet (emails, instant messaging, online telephone, etc.) helps immigrants to keep up existing ties in the homeland. Moreover, immigrants often have to build a new network from scratch in the host country. Facilitating easier and faster access to more diverse social milieus and broader ranges of information and resources, the Internet is instrumental in building and maintaining such transnational networks (Wellman 2002). By affording changes in the pattern, reachability, and speed of networking, the Internet affects the structure and composition of the networks of immigrant entrepreneurs. A biotech research scientist, BPS recalled how she became connected to a professional association and eventually to the Zhongguancun Science Park, the flagship of state-sponsored transnationalism to attract overseas Chinese to start up high-tech firms in China.

> It was just my first year in Toronto. Except for a few friends at the church, I didn't have many social contacts. I saw the news about the inception of the Canadian Chinese Biotech Association on the web and sent them an email. Soon, they held the opening ceremony. I met a lot of biotech professionals on

> the spot. Several days later, I got an email from the association, calling people to contact the liaison office of Zhongguancun Science Park in Toronto. They were organizing a tour for overseas Chinese who wanted to start up in their incubators. As I had already booked tickets for China..., it was convenient for me to attend. I sent the organizer an email and registered for the tour (2003, Beijing).

BML, the owner of a small Internet telephony business in Toronto, provided a similar account of how the Internet helped him to reconnect with a lost contact in the new country.

> I posted a message on China Smile, asking if there was anyone providing Internet telephony service to China. Three hours later, the post was answered by a guy who left his telephone number and invited me to call. I called. Guess what? He was one of my clients when I worked for Siemens China six years ago... I cannot stop thinking that there are so many crouching tigers and hidden dragons in Toronto's Chinese community. And you can reach them through the Internet. I do not know why such a highflier like him migrated. Anyway, he is my biggest supplier now (2003, Toronto).

Once logged onto the Internet, immigrant entrepreneurs can tap into previously unattainable resources. A border-crossing network helps entrepreneurs dispersed in different geographic locations "optimally combine and integrate resources." As illustrated in the following quotation of BML, the Internet helps immigrant entrepreneurs to pool resources embedded in their networks, which are often scattered in different places along their immigration trajectory:

> We began our operation as a guerilla war. There are three of us. The tech guy is in Boston, the sales guy is in Vancouver, while I am the guy for everything else and based in Toronto. We are lucky that we live in the Internet era. The Internet makes it handy to put dispersed resources together. We keep in close contact with one other through instant messaging, email, phone, and fax (2004, Toronto).

Accelerating the global flow of information, the Internet has the potential to transcend geographic boundaries. It enables immigrants to create new ties with people of shared interests outside of their often-restricted physical world, and thus gain access to resources beyond their local setting. This is especially important to immigrants who lack access to mainstream banks or venture capitalists (VC) to gain startup capital. For example, SZX and his co-founders were desperately searching for seed capital. On the one hand, the team was concerned that entering into a joint venture with companies in China would make the joint venture too difficult to control. On the other hand, it had been difficult to find Canadian VCs and bankers. Touring around in various trade shows and expos in China to promote their project did not deliver any immediate results either. Then, one day, "A Hong Kong VC firm contacted us. The boss is a friend of Li Ka-Shing. They saw our environmentally friendly equipment on the website of the high-tech expo. The Internet is powerful. For the first time, money is chasing us..." (2003, Toronto).

E-commerce

The Internet is not just a medium of communication. Immigrant entrepreneurs use it as a tool for marketing and reaching out customers. JC, who has an MBA from a top Canadian business school and runs a multi-level marketing business in Toronto and Shanghai, wrote to his downlines in an email:

> I have set up a website to develop it as e-commerce. I do a lot of online marketing. The company provides an e-business package. Newsletters and advertisements are sent to potential customers automatically... I was working during Canada Day on a very exciting project for our team. I am building a system for every one of us in the team to generate quality leads and recruit them from anywhere in the world 24/7, especially into U.S. market where the... business is getting hot. This means that you can leverage both online and offline technology to do the business even while you are sleeping. The system will pre-screen people for you and get them to come up to you to sign up. I can help you to set up your prospecting site in 30 minutes. So, please check my prototype right here... (2003, Toronto).

Interestingly, immigrant entrepreneurs do not passively wait for people to come and visit their websites. Instead, they are proactive in grabbing the attention of potential customers. Understanding that search engines play a gate-keeping role in channeling information to users, they deliberately design their websites to improve rankings in Google or other search engines. In the word of KZ, "the Internet really helps a lot" to enhance his company's visibility:

> We made tremendous effort to set up our website. It has been bilingual, Chinese and English, right from the start. Our website caters to Google's searching technologies. We don't pay [to increase our ranking in Google]. However, if you google each of our core products, we always show up in the front page. In fact, most clients find us through the web (2003, Toronto).

The Internet is Not Everything — Media Multiplexity and the Dynamic of Online and Offline

The Internet is critical to entrepreneurship. However, immigrant entrepreneurs are aware what the Internet can and cannot do. The Internet is not a silver bullet that can solve all problems. The Internet is embedded in the broader context of media use. It is just one among many communication media. While it creates a new social space that restructures existing relations and encourages the emergence of new social practices, the online and the offline are intertwined with each other. Cyberspace is not an isolated place separated from real life.

Immigrants use multiple forms of communications media, in accordance with whichever is most convenient and socially appropriate. Instead of substituting existing ways of keeping in contact, the Internet supplements face-to-face visits and telephone contacts. For instance, JC may sound over-enthusiastic about his online marketing system; however, he warned his downlines in the same email:

> This system is only the add-on and it can never replace picking up your phone and talking warmly to people and inviting them for a one-on-one or two-on-one meeting. That's by far the most effective. The only difference is that you will have the leverage of the technology to work for you while you sleep and get into the other markets when you cannot physically go there to meet someone (2003, Toronto).

Sometimes, the lack of physical infrastructure prohibits the use of the Internet, as revealed in the following quotation. SH landed in Canada in the late 1990s. Graduating from a top Canadian business school, he first set up an e-commerce company in Toronto, which never took off and eventually closed down in 2001. Soon, he put together a team of Chinese-Canadian engineers and professionals, returned to Beijing, and founded a software firm that specializes in providing customized enterprise management systems to big energy companies.

> We have always been using the Internet. However, we use the Internet less and less. Because the big chuck of communication in our firm occurs between the head office and clients' project sites. When we enter the site, it is often the case that they are still excavating or draining the swamps. There is no Internet. The most viable tool of communication is the cell phone. We communicate a lot through cell phone. All core team members have a big cell phone bill every month. We use SMS a lot... (Beijing, 2004).

The social fabric also matters. In order to add a human touch to their transnational network, immigrant entrepreneurs travel regularly between China and North America. Some travel an average of seven or eight times a year. Others travel less often but stay longer, as they have found that "staying a couple of days just doesn't work." Each visit is packed with dining, drinking and entertaining business partners, clients, friends, and families. For example, JL talked about how, during his three-week visit to China in 2003, his mom was upset as he was busy catching up with business friends and did not eat breakfast at home even once. In the following quotation, YF offers a similar account. To warm up and reactivate the network, he endured many banquets and karaoke sessions:

> Every day during the two months [in Shanghai], there was dinner. Partners in China told me that "your stomach is not your stomach. Your stomach is the company's stomach. You eat and drink for the company." Well, I am not sure if the company will feel my stomachache. Yet, you want to keep people happy. It is easier to get things done when people are in a good mood. If everyone wants to have some fun and you show a face of "class struggle," you offend people. Sometimes, it seems that there is no chance of cooperation. However, after several karaoke sessions and bottles of wine, things take a dramatic turn... (2003, Toronto).

Discussion — the Digital Divide and Network Diversity

As Guarnizo *et al.* commented, participating in transnational activities is "associated with greater stability and greater resources brought from the home country" (2003:1232). With higher human capital, larger networks, and legal citizenship status in the country of settlement, transnational entrepreneurs are the better-off segment in the immigrant community (Landolt 2001). While almost all interviewees I met turned out to be Internet users and found the Internet to be an important tool to communicate, search for information, and keep in contact with customers and friends, there is also evidence to suggest the existence of a digital divide based on age. For example, LC operates an advertising agency. He tried to persuade his uncle who runs a summer resort to go online:

> Many Chinese old-timers often have no idea about "marketing." They lack that kind of mentality... I told him that he needed a website. It only costs him $500. He said "No, no, noooooooo. Why spend $500? My son can do it. He is in high school and just learned how to build a homepage." (2003, Toronto).

Another question that remains open is the question of to what extent the Internet increases network diversity. A new kind of homophily may emerge as people now have greater opportunities to find like-minded people on the web (McPherson *et al.* 2001). One study shows that Koreans and Chinese in Los Angeles make new connections on the Internet mostly within their own ethnic groups (Matie and Ball-Rokeach 2002). As the webmaster of a Toronto-based website pointed out, Chinese immigrants often surf different websites based on their places of origin, socioeconomic backgrounds, and extent of integration into Canadian mainstream society (Zhao 2004). While the Internet affords communications that transcend geographic proximity, to what extent does it help entrepreneurs cross ethnic lines and find ties outside of their own ethnic group?

Conclusion — This is Just the Beginning

Transnational entrepreneurship offers a unique vantage point to understand the dynamic interaction of technology, agency, culture, and social networks, involving the actors' motivations and choices, network formation, institutional reconfiguration, and technological transformation. It also provides the opportunity to examine the relationship between Internet use and critical social outcomes in specific institutional, organizational, and cultural contexts. The existing literature has attributed the rise of the current wave of transnationalism to the use of new technologies. However, there has been little systematic research on how actors in the transnational field actually use new technologies to participate in transnational activities.

This chapter is one of the first to discuss the socio-technological dimension of transnational entrepreneurship. Drawing on in-depth interviews and fieldwork conducted in both the country of settlement and origin, the chapter systematically explores how immigrant entrepreneurs rely on communication networks and social networks to engage in transnational business activities.

State policies have resulted in an increase in the Chinese immigrant population in major Canadian cities, especially that of skilled and business immigrants from mainland China. This growing immigrant population is providing the market, labor, and capital for immigrant entrepreneurship. For immigrants who are facing great difficulties in the labor market, entrepreneurship offers an alternative approach to economic integration. Moreover, Chinese immigrants use the Internet and other communication technologies intensively. The diffusion of the Internet in China is making it possible for immigrants to keep in contact with their friends and families in China through the

Internet. Furthermore, social, cultural, and geographical distances between North America and China are creating opportunities for Chinese immigrant entrepreneurs to act as brokers.

The data has shed light on the importance of the Internet to transnational entrepreneurship. To participate in transnational entrepreneurial activities, immigrant entrepreneurs use the Internet and other ICTs to seek information, develop social networks, and mobilize and integrate resources. The Internet is instrumental in building and maintaining border-crossing social networks. It helps immigrant entrepreneurs enhance ties that bond and increase ties that bridge. It also helps them to nurture and catch up with cultural capital. However, the Internet is not everything. Transnational entrepreneurs selectively use multiple forms of communication media, in accordance with whichever is most convenient and socially appropriate — email, telephone, and face-to-face interactions — to keep in contact with family, friends, workmates, suppliers, and customers.

Will the current wave of transnational entrepreneurship continue? The answer is — this is just the beginning. Chinese transnational entrepreneurship is growing along with the immigrant population and China's increasingly important role in the global economy, both as the world's factory and an emerging market with a huge demand for goods and services. Accordingly, a relatively high proportion of Chinese immigrants have already been involved in transnational economic activities. For instance, in Australia, Lever-Tracy and Ip found that "even small businesspeople are weaving global strategies that may or may not be fanciful" as one-third of the Chinese small businesses they interviewed were involved in transnational entrepreneurship (1996:268). If the current trends continue, Chinese transnational entrepreneurship — afforded by the Internet and embedded in border-crossing networks — will follow an upward curve in the near future.

The chapter has at least two limitations. First, it has focused on how actors in the transnational field use the Internet to engage in entrepreneurship. However, there might be immigrant entrepreneurs who have expanded their business transnationally without using the Internet. The discussion about the effect of the Internet on transnationalism needs to incorporate both the impact of the Internet on those who can afford to access and use it and on those who cannot. Second, the data presented here are based on interviews and participant observations, which as Portes and colleagues have pointed out (1999), inevitably samples on the dependent variable and thus may overestimate the impact of the Internet on transnational entrepreneurship. Working with Prof. Wellman, I am conducting a survey of transnational entrepreneurs in Toronto's Chinese community. The survey will allow us to compare different patterns of use of the Internet and other ICTs by immigrants; to measure the characteristics of transnational entrepreneurial networks; and analyze how communication networks, mediated by social networks, affect the resource mobilization and business success of transnational entrepreneurs.

I express my appreciation for the financial support provided by the Social Sciences and Humanities Research Council of Canada, the Vivienne Poy Chancellor's Fellowship in the Humanities and Social Sciences, and the Dr. David Chu Fellowship in Asian Pacific Studies. I am thankful to the following for their comments: Barry Wellman, colleagues at the NetLab, Adrian Favell, and participants of the UCLA/SSRC Summer Institute on International Migration in June 2004. The opinions expressed here are solely my own.

References

ACNielsen. 1999. "Toronto Chinese Media Index." Toronto.

Agre, P. E. 2002. "Real-Time Politics: The Internet and the Political Process." *The Information Society* 18:311–331.

Basch, L., N. Glick-Schiller and C. Szanton-Blanc. 1994. *Nations Unbound: Transnational Projects, Postcolonial Predicaments and Deterritorialized Nation-States.* Amsterdam: Gordon and Breach.

Calhoun, C. 1998. "Community without Propinquity Revisited: Communications Technology and the Transformation of the Urban Public Sphere." *Sociological Inquiry* 68:373–397.

Castells, M. 1996. *The Rise of the Network Society.* Malden, MA: Blackwell.

Chen, W. and B. Wellman. (Forthcoming.) "Fathoming the Digital Divide." In *Blackwell Companion to Social Inequality*, edited by M. Romero and E. Margolis. Oxford: Blackwell.

China Internet Network Information Center (CNNIC). 2004. Semiannual Survey Report on the Development of China's Internet. http://www.cnnic.org.cn/

DiMaggio, P., E. Hargittai, W. R. Neuman and J. Robinson. 2001. "Social Implications of the Internet." *Annual Review of Sociology* 27:287–305.

Emirbayer, M. and A. Mische. 1998. "What is Agency?" *American Journal of Sociology* 103(4):962–1023.

Fong, E. and E. Ooka. 2002. "The Social Consequences of Participating in Ethnic Economy." *International Migration Review* 36:125–146.

Georgiou, M. 2003. "Mapping Diasporic Media Across the EU: Addressing Cultural Exclusion." London: Report for the European Commission within the EMTEL II Network.

Glick-Schiller, N., L. Basch and C. Blanc-Szanton, eds. 1992. *Towards a Transnational Perspective on Migration: Race, Class, Ethnicity, and Nationalism Reconsidered.* New York, NY: New York Academy of Sciences.

Guarnizo, L. E., A. Portes and W. Haller. 2003. "Assimilation and Transnationalism: Determinants of Transnational Political Action among Contemporary Migrants." *American Journal of Sociology* 108:1211–1248.

Guo, L. 2002. "Frequency Questionnaire of the Survey on Internet Usage and Impact." Beijing: Research Center for Social Development, Chinese Academy of Social Sciences.

Hargittai, E. 2002. "Second-level Digital Divide: Differences in People's Online Skills." *First Monday,* 7(4). http://firstmonday.org/issues/issue7_4/hargittai/index.html

Haythornthwaite, C. and B. Wellman. 2002. "The Internet in Everyday Life: An Introduction." In *The Internet in Everyday Life*, edited by B. Wellman and C. Haythornthwaite. Oxford: Blackwell.

Haythornthwaite, C. and C. Hagar. (Forthcoming.) "The Social World(s) of the Web." *Annual Review of Information Science and Technology.*

Haythornthwaite, C. 2001. "Exploring Multiplexity: Social Network Structures in a Computer-supported Distance Learning Class." *Information Society* 17(3): 211–226.

Hiebert, D. and D. Ley. 2003. "Characteristics of Immigrant Transnationalism in Vancouver." Vancouver Centre of Excellence: Research on Immigration and Integration in the Metropolis.

Howard, P. N. (Forthcoming.) "Embedded Media: Who We Know, What We Know, and Society Online." In *Society Online: The Internet in Context*, edited by P. N. Howard and S. Jones: Sage Press.

Katz, J. E. and R. E. Rice. 2002. *Social Consequences of Internet Use: Access, Involvement, and Interaction.* Cambridge, MA: MIT Press.

Landolt, P. 2001. "Salvadoran Economic Transnationalism: Embedded Strategies for Household Maintenance, Immigrant Incorporation, and Entrepreneurial Expansion." *Global Networks* 1:217–241.

LaRose, S., F. Guay and M. Boivin. 2002. "Attachment, Social Support, and Loneliness in Young Adulthood: A Test of Two Models." *Personality and Social Psychology Bulletin* 28:684–693.

Lever-Tracy, C. and D. Ip. 1996. "Diaspora Capitalism and the Homeland: Australian Chinese Networks into China." *Diaspora* 5:239–274.

Levitt, P., J. DeWind and S. Vertovec. 2003. "Transnational Migration: International Perspectives." *Special Issue of International Migration Review* 37.

Levitt, P. 2001. "Transnational Migration: Taking Stock and Future Directions." *Global Networks* 1:195–216.

Li, P. 1993. "Chinese Investment and Business in Canada: Ethnic Entrepreneurship Reconsidered." *Pacific Affairs* 66:219–243.

Li, P. 2002. "Economic Return of Immigrants' Self-employment." Ottawa: Citizenship and Immigration Canada.

Matei, S. and S. J. Ball-Rokeach. 2002. "Belonging in Geographic, Ethnic, and Internet Spaces." In *The Internet in Everyday Life*, edited by B. Wellman and C. Haythornthwaite. Oxford: Blackwell.

McPherson, M., L. Smith-Lovin and J. M. Cook. 2001. "Birds of a Feather: Homophily in Social Networks." *Annual Review of Sociology* 27:415–444.

Menzies, T. V., G. A. Brenner, L. J. Filion, L. Lowry, C. Perreault and C. Ramangalahy. 2000. "Transnational Entrepreneurship and Bootstrap Capitalism: Social Capital, Network and Ethnic Minority Entrepreneurs." In *The Second Biennial McGill Conference on International Entrepreneurship*. Montreal, Canada.

Miller, D. and D. Slater. 2000. *The Internet: An Ethnographic Approach*. Oxford: Berg.

Mitra, A. 2001. "Diasporic Voices in Cyberspace." *New Media and Society* 3:29–48.

Miyata, K. 2002. "Social Support for Japanese Mothers Online and Offline." In *The Internet in Everyday Life*, edited by B. Wellman and C. Haythornthwaite. Oxford: Blackwell.

Ong, A. and D. Nonini, eds. 1997. *Ungrounded Empires: The Cultural Politics of Modern Chinese Transnationalism*. New York: Routledge.

Panagakos, A. 2003. "Downloading New Identities: Ethnicity and Technology in the Global Greek Village." *Identities: Global Studies in Culture and Power* 10: 201–219.

Portes, A., ed. 1995.*The Economic Sociology of Immigration: Essays on Networks, Ethnicity, and Entrepreneurship*. New York: Russel Sage Foundation.

Portes, A., L. E. Guarnizo and P. Landolt, eds. 1999. *Special Volume on Transnational Migrant Communities, Ethnic and Racial Studies* 22(2).

Portes, A., W. J. Haller and L. E. Guarnizo. 2002. "Transnational Entrepreneurs: An Alternative Form of Immigrant Economic Adaptation." *American Sociological Review* 67:278–298.

Portes, A. 2003. "Conclusion: Theoretical Convergencies and Empirical Evidence in the Study of Immigrant Transnationalism." *International Migration Review* 37:874–892.

Quan-Haase, A., B. Wellman, J. Witte and K. Hampton. 2002. "Capitalizing on the Internet: Network Capital, Participatory Capital, and Sense of Community." In *The Internet in Everyday Life*, edited by B. Wellman and C. Haythornthwaite: Oxford: Blackwell.

Reitz, J. G. 2001. "Immigrant Skill Utilization in the Canadian Labour Market: Implications of Human Capital Research." *Journal of International Migration and Integration* 2:347–378.

Saxenian, A. L. 1999. *Silicon Valley's New Immigrant Entrepreneurs*. San Francisco, CA: Public Policy Institute of California.

Saxenian, A. L. 2002. *Local and Global Networks of Immigrant Professionals in Silicon Valley*. San Francisco, CA: Public Policy Institute of California.

Smith, M. P. and L. E. Guarnizo, eds. 1998. *Transnationalism from Below, Comparative Urban and Community Research*, Volume 6 New Brunswick: Transaction Publishers.

Smith, R. C. 1998. "Transnational Localities: Community, Technology and the Politics of Membership within the Context of Mexico and U.S. Migration." In *Transnationalism from Below, Comparative Urban and Community Research*, Volume 6, edited by M. P. Smith and L. E. Guarnizo. New Brunswick: Transaction Publishers.

Spooner, T. 2001. *Asian-Americans and the Internet: The Young and the Connected Pew Internet and American Life Project*. Washington, DC. Available from http://www.pewinternet.org/

Vertovec, S. 2003. "Migration and Other Modes of Transnationalism: Towards Conceptual Cross-Fertilization." *International Migration Review* 37:641–665.

Waldinger, R. and D. Fitzgerald. 2004. "Transnationalism in Question." *American Journal of Sociology* 109:177–195.

Wellman, B. 2001. "Physical Place and Cyber-Place: Changing Portals and the Rise of Networked Individualism." *International Journal for Urban and Regional Research* 25:227–252.

Wellman, B. 2002. "Little Boxes, Glocalization and Networked Individualism." In *Digital Cities II: Computational and Sociological Approaches*, edited by M. Tanabe, P. van den Besselaar, and T. Ishida. Berlin: Springer.

World Economic Forum. 2002. *Annual Report of the Global Digital Divide Initiative*. Geneva: World Economic Forum.

Yang, G. 2003. "The Internet and the Rise of a Transnational Chinese Cultural Sphere." *Media Culture & Society* 25:469–490.

Zhao, P. 2004. *Are There Three Different Cultural Circles among Chinese Canadians?*, cited December 30th, 2004. Available from http://www.51.ca/news/s/2004-01-22/011389.shtml.

Zhou, M. and G. Cai. 2002. "Chinese Language Media in the United States: Immigration and Assimilation in American Life." *Qualitative Sociology* 25:419–441.

Zhou, Y. and Y.-F. Tseng. 2001. "Regrounding the 'Underground Empires': Localization as the Geographical Catalyst for Transnationalism." *Global Networks* 1:131–153.

10 *Mobile Phones and New Migrant Workers in a South China Village: An Initial Analysis of the Interplay between the "Social" and the "Technological"*[1]

Wai-chi Chu and Shanhua Yang

Introduction

This study aims to depict how the cell phone, as one type of information and communication technology (ICT), is interfacing with the cultural traditions of peasant migrants working in township enterprises in the city of Dongguan in Guangdong Province in southern China. Although this is a substantive piece

[1]This paper is an updated and revised version of the conference paper titled "Cell phone in use: The touch of technology as experienced by migrant workers in a southern Chinese Town" that was presented at the 36th IIS World Congress held in Beijing, the People's Republic of China between July 8th to 12th under the Session "The Co-construction of Society and Technology in a Global World" chaired by Leopoldina Fortunati. This article is one of several research papers arising from two research grants ("Project on Cultural Hermeneutics" and "An Exploratory Study on Mobile Telephony and the Social Lives of Outside Workers in Southern China") supported by a research grant from The Hong Kong Polytechnic University. Besides the two authors, the research collaborators include Law Pui-lam from Department of Applied Social Sciences of The Hong Kong Polytechnic University, Cheng Weimin from Department of Sociology of Peking University, Liu Xiaojing from the Rural Development Institute of the Chinese Academy of Social Sciences of the People's Republic of China, and two research students, Peng Yinni and Song Jing, from the Department of Sociology of Peking University. The interviews reviewed in this article were based on fieldwork carried out in Tangxia Town of Dongguan City in Guangdong Province between July 2003 and March 2004. A total of 20 migrant workers (13 males and 7 females) were interviewed in 2003, and 18 migrant workers (7 males and 11 females) were interviewed in 2004. Of the latter group, three had previously been interviewed in 2003. The statistics adopted in the three appendices are excerpted from different monthly figures from the *Tongji Yuebao* (*Statistics Monthly*) of the Ministry of Information Industry (MII) of the People's Republic of China.

of research, there are at least two issues that the authors will address in hopes of opening up further discussions in this area of research. The first is that we observe that there is a gap in research on the use of cell phones in China. As will be seen from the data provided in this discussion, the development of the cell phone market in China is a subject deserving of much more sociological discussion than is the case at present.[2] The reason why the subject has not yet received the attention it deserves may be because the cell phone is only one kind of ICT. Since the notion of "information society" was raised by Daniel Bell in the 1970s, sociological literature on the consequences to human life of the emergence and expansion of ICTs has placed more emphasis on the network aspect as derived from the development of the Internet, than on other aspects of the issue. This is the case even with the leading figures in the field, such as Frank Webster in the 1980s or Manual Castells in the 1990s.[3] By the same token, there are studies that discuss the Internet or cyber development in China,[4] but sociological literature in the English-speaking world relating to the use of cell phones in China is scant. Second, we want to contribute to meta-theoretical reflections in this area of research. This is of particular importance when ICTs have a global dimension by default, and their impact on the human race can hardly be avoided. Thus, a cross-cultural comparison and debate on the subject is necessary. Here, we are not only concerned about whether or not analyses of the impact of an information society on the human race are universally applicable.[5] We are also concerned about the meaning of "understanding" in this kind of cross-cultural research, where a Eurocentric and/or ethnocentric attitude may unconsciously creep into the analysis of the researcher, thus distorting the observations and analysis. If we do believe that there is a constant interplay between "technology" and "society" instead

[2]According to our literature search, thus far there is only one article that deals directly with the use of cell phones in mainland China; see Yu and Tng (2003). Yet this article is largely a discussion of culture and the design of cell phones for China, and focuses on the use of cell phone in urban China, with no mention of the rural aspect of the issue.

[3]For instance, Gaser (2004) pointed out that in the first of the three-volume work by Manuel Castells (1996), only the Internet is given status, while mobile communication facilities are almost totally neglected. Not only did Castells continue to pay little attention to mobile telephony in his later work (2004), Webster (2002, 2004) has also shown no interest in including this communication device in the main body of his work.

[4]The study of Qui (2004) is a case in point. One can see from the reference list at the end of the discussion that there have been a large number of studies in this area, but there are no references to any sociological discussions of mobile telephony in mainland China.

[5]For instance, there are four related claims in most versions of the idea of an information society, namely, that it involves a social revolution, a transformation of work, a new kind of politics, and the decline of the state (May 2002).

of regarding technology as somewhat "outside" of society and impinging upon it (as is suggested by the use of such a phrase as the "social impact of technology,"[6]) we must accommodate the cultural traits of a context to act and react with ICTs. In fact, as we are going to show below, there is particular cultural meaning in the use of cell phone to peasant workers who have their own unique background that is so different from that of their predecessors, thus perhaps opening up some interesting insights for more fruitful cross-cultural comparisons in the future.

The Research Question

What kinds of challenges to sociological theory and research are posed by the expansion of cell phones/mobile cellular phones/mobile telephony usage worldwide? While the cell phone seems to have evoked much less intellectual enthusiasm and scientific research than the World Wide Web, a much broader strata of the world's population is using cell phones than are using personal computers and other net technologies. Simply put, an uneducated person may not be able to use a computer, but as long as one knows how to press numbered buttons, one may use a cell phone.[7]

There have been discussions acknowledging the sociological implications on human life since the inception of the telephone.[8] Simply put, the fixed phone broke the constraints of physical proximity on human beings, but still keeps people in a certain space. With the emergence of wireless transmission technologies, which then developed into ICTs and then to the birth of cell phones, people have been brought into a new era where they can engage in communication that is free from the constraints of physical proximity *and* spatial immobility. Or, phrased in another way, the spatial and temporal limitations on human interaction have been reduced. At once, the sociological meaning of this drastic change on human interaction becomes an interesting issue. We are talking about new ways of life and new identities, from the point

[6]See the discussion of Lyon (1988), who devotes his whole book to arguing against the views of the "social impact of technology."

[7]See Townsend (2000) for a supportive argument on the penetration of cell phones to the illiterate mass population. Certainly, whether the fact that a person is limited by illiteracy to the use of the cell phone represents another kind of digital divide is a subject that deserves a lengthy discussion, but this does not fall within the scope of this discussion. The point here is that the cell phone per se as a product of the information society is less difficult for a person to use than a computer, let alone the Internet, which requires a great deal in terms of language management skills.

[8]On the sociological meaning of fixed phones, see the discussions of Aronson (1971), De sola Pool (1977), Dimmick *et al.* (1994).

when the cell phone was seen as an elite device (Ross 1993), to the fact that it is creating a new aspect of life in which all human beings are equal, irrespective of age, gender, wealth, or cultural background (Gaser 2004); from its possession of a communicative function to the achievement of an ice breaking time-space distanciation to the articulation of the enjoyment of "civil inattention" — or the privatizing of public spaces — by its use in front of strangers (Cooper 2000; Puro 2002);[9] from becoming a kind of "vehicle of speed" of the users (Kopomaa 2000) to allowing the users a kind of "Martini benefit" (Fox 2001). At a glance, we are achieving a kind of "emancipation" of the entire human race, but does the use of cell phone have no drawbacks?

Rather than leading to a tendency to open up to new acquaintances, using a cell phone may also support tendencies towards social closure. Certainly, the cell phone can strengthen networks, but from another perspective, it can also weaken communities (Kopomaa 2000; Wellman 2001). That is to say, by celebrating its ability to sustain a perpetual contact, one may easily overlook the very fact that using a cell phone is also a way to hide from a time-space to which one does not feel much of a sense of belonging. Certainly, when our interpersonal and institutional relationships are increasingly "virtual" and "mobile," we have to face squarely the concern of whether or not technologies have transformed the meaning of "community," and that is why mobile technologies present particular kinds of interpersonal and organizational uncertainties (Green 2003). This forces us to rethink the *further* meaning of "trust" and "risk," an issue raised so succinctly by renowned figures such as Beck (1992) and Giddens (1990) in the early 1990s, when the cell phone was still at the dawn of its development.

Needless to say, then, ICTs are both empowering and controlling. The possibility that "you cannot be found" when you switch off your cell phone is functioning side by side with the possibility of "forced accessibility" once you do have a cell phone (Puro 2002; Kopomaa 2000). The crucial point of concern here is that the cell phone is perhaps the most influential late modern instrument that blurs the spheres of public and private (Fortunati 2002). At the end of the day, the exploration of whether cell phone usage is bringing us to a new level of empowerment or a new form of slavery is, perhaps, the very old sociological concern of "who, at what time, owns what forms of power." It is thus not surprising to predict that, sooner or later, there will be a trend in

[9]Using a cell phone can convey the message that the user is physically, but not spatially alone, as the machine can link a person up with his/her preferred embedded social setting.

academia of moving from "a sociological theory of the mobile phone" to "a sociological theory of the caller identification function of the mobile phone!"

What is the purpose of outlining the above research awareness in the field? It is a fact that the market for telephones in mainland China is sky-rocketing. Appendix I of this chapter earmarks two periods, July 2003 and December 2004 for comparison.[10] In July 2003, the total number of users of fixed phones was 240.7 million, and that of cell phones was 239.4 million. By December 2004, when the number of fixed phone users had risen to 312.4 million (an increase of nearly 30%), there were 334.8 million users of mobile phones (an increase of nearly 40%). As for the penetration rate, Appendix II shows that whereas that of fixed phones increased from 17.5/100 to 24.9/100 (an increase of around 40%) in a two-year period, the penetration rate of mobile phones was initially lower but overtook that of fixed phones: from 16.2/100 to 25.9/100 (a nearly 60% increase). This in itself shows the value of doing research on mobile telephony in China. As for why we have chosen to focus on Guangdong Province, as stated in Appendix I, in both periods, Guangdong had the highest number of fixed and mobile phone users. In December 2004, whereas the number of fixed phone users had reached 27.59 million and that of cell phone users 44.97 million, the figures 16 months ago were only 22.2 million and 36.45 million, respectively (Law 2003). While the percentage increase within this period in Guangdong Province was not the highest in the nation, an increase of nearly 50% from such a huge base is itself a strong reason to make Guangdong Province the focus of our study.

There are more reasons why the subject is worth studying. How can we explain the sociological and cultural meanings of the phenomenon of new migrant workers spending, on average, three to four months of their monthly income on buying cell phones, and that such phones are often more expensive models than those used by the entrepreneurs that employ them? What is the function of the cell phone to these workers? Why have they chosen the cell phone instead of some other consumer item available in the market? Do the answers to the above questions echo the findings outlined above in the Western literature, or do they shed new light on the argument that, given the almost ubiquitous adoption of cell phones within and across current human societies

[10]July 2003 marked the period of which the data of the first paper of our research team (Law 2003) was based on, and December 2004 was the time when the latest statistics provided by Ministry of Information Industry (MII) was available on its website as the draft of this paper was finalized.

and cultures, it is dangerous to assume that there is no heterogeneous mobile effect?

The Touch of Cyberspace by the New Migrant Workers

Foreign-invested and local-invested enterprises have emerged in Tangxia Town of Dongguan City in the last two decades just as in many other towns in Guangdong Province since the 1970s, in response to the open door policy. One of the 32 towns under the administration of Dongguan, Tangxia Town is comprised of 20 villages, with a total of around 33,000 local residents and over 400,000 workers from outside the town (Law 2003). But this cohort of migrant workers is very different from that of the first generation of migrant workers that appeared at the start of the open-door policy. In recent years, scholars in mainland China have focused an increasing amount of attention on the emergence of a new group of workers that have been migrating to the cities. This group of people deserves attention because they have their own reasons for going off to work, and their presence may give rise to new issues in the "city-village merger" effort (Luo and Wang 2003; Wang 2001). Here, we try to summarize the characteristics of this group of people by, first, linking the issue to the boarder socio-economic background of the Pearl River Delta Region, and second, by analyzing the perceptions they have of others and themselves against such a socio-economic backdrop. Simply put, from the early 1980s to the turn of the century, manufacturing industries, characterized by their labor-intensive nature, are gradually being replaced by the electronics and information industries. Apart from being less labor intensive, such industries place more demands on workers in terms of knowledge and technical skills. Those with low — or even no — educational qualifications, those with less working initiative, or those older people, all become less attractive in times of change, and entrepreneurs are more inclined to replace them with younger people from villages. Yet the new migrant workers have their own socio-cultural backgrounds that bring new problems for their employers and for themselves as well. With the development of China, this group of young migrant workers has been able to achieve at least some junior secondary education and, through access to television broadcasts, they have been exposed to city culture long before they enter a city. Even though they are still quite deprived, their parents, who were the first generation of migrant workers to cities, have tried to protect them from living a life as difficult as the one they themselves experienced. This desire for upward mobility has affected this new generation in several ways: first, they have not had much experience of the difficult life of peasants which they, by default, dislike; second, they unfortunately do not have the means to lead a real city life, although they aspire to do so. In other words, unlike their predecessors, who worked in the city for economic gain, and thus were willing to endure a hard life for a "grand return" to their home town, from the very start this group of new migrant workers regard working in the city as a kind of adventure. That is to say, they feel that working in a township enterprise will give them the chance to have a feel for city life. The problem, of course is that they are in a much better position than their seniors to realize the limits of having a peasant identity in a rapidly developing Chinese economy, but do not see any way of shaking off this identity. Thus, they are much more frustrated than the first generation. This is where the focus of our research lies: the popularity of cell phones within this group of new migrant workers must be understood against the particular socio-economic predicament in which they find themselves.[11]

We have looked for an explanation of this popularity from the beginning of the interviews we conducted in Tongxi Town. In the early 1990s, the average salary for migrant workers was about 200 to 400 yuan; it is now uplifted to around 400 to 600 yuan. At the turn of the last century, cell phones began entering the market. Before long, because of severe market competition, the price of cell phones started dropping dramatically. For instance, a Nokia 8310 was 2,500 yuan in 2002, 1,400 yuan in 2003, and by June 2004, had dropped to below 1,000 yuan. At the same time, the expectations of the workers kept growing. The Nokia 8310 only has a black-and-white display, while the migrant workers regard color display as a must. Yet, getting a color display cell phone would cost at least three months of the average salary of a migrant worker. What is more, once the workers have become accustomed to using a cell phone, it seems that cheap rates will only turn them into more frequent users.

Apparently, on top of voice calls, the sending of SMS messages is becoming more commonplace. According to statistics, 18.9 billion messages were sent in 2001, but the figure shot up to 90 billion in 2002, for an average of 247 million messages a day.[12] But such an increase means nothing when compared

[11]The popularity of cell phones was clear from the information we collected on February 28th, 2004. For example, informants YZY, LM, and TJR told us that " ... basically we all have a cell phone," with "we" referring to 6,000 workers in a large electronic device factory in Tongxia Town. In another interview conducted on the same day with two workers (GF and WK) in a golf factory, we were told that in the dormitory unit provided by the factory, where the two interviewees were living, only two out of the six residents did not have a cell phone.

[12]" 'Thumb Economy' earned 9 Billion", *Mingpao Daily*, September 14th, 2003, B6.

with the figures in Appendix III. A glimpse at the figures shows that the number of SMS messages sent by mobile phone skyrocketed in 2004 along. The data is astonishing: while the number of mobile phone users at the end of the year end had increased by around 21% from the start of the year, the increase in of the average number of SMS messages sent on each mobile phone increased by ten times more (or an increase of more than 1,000%)! Perhaps the explanation is that the sending of SMS messages is a less expensive way of maintaining virtual contact. Also the technological upgrade in the period has been so rapid that the number of Chinese characters that could be contained in each message shot up from 200 to 400 in a year's time (Law 2003). As a matter of fact, according to the service package of May 2004 as offered by China Mobile — the largest network provider in mainland China, the charge for each message was 0.15 yuan; and that of China Unicom, the next largest service provider, was only 0.1 yuan. In mid-2004, China Mobile promoted a new storage card known as M-Zone, which charged 26 yuan a month for 300 messages. This was very well received by the migrant workers we met. But again, cheap rates do not mean that the workers are spending less, as cheap rates can easily generate more usage. In fact, our interviews with the workers were almost invariably interrupted by incoming SMS messages for them.

An interesting question arises: of the many advanced consumer durables (lap top computers, MP3 players, digital cameras, cell phones, and so forth) that the new migrant workers come across, why have so many of them chosen the cell phone? One obvious answer is that, because they are working in a "foreign" place, a cell phone allows them to keep in contact with their families easily. Indeed, in some of our early interviews, one reason always given for buying cell phone was that it made them accessible.[13] Relying on the fixed phone line provided by the factory dormitory was hardly an alternative, as the phone could only be used after office hours, a time when it was in great demand. Certainly, one could use a public phone, but, as stated by LZB, "... you have no desire to go out once you have had your bath." But once a cell phone is consumed, a different story emerges. We found that workers are, on the whole, very cautious about using cell phones for long distance calls. A cell phone is used to notify the other party with a certain kind of ring code or simple conversation to expect a call. What follows is the termination of the wireless conversation, followed by a fixed phone call to complete the conversation. It seems that once a cell phone is purchased, the original

[13]Interview with informants LZB and YWM in the morning session of July 21st, 2003, as well as that with ZXX in the morning session on July 22nd, 2003.

intention of providing easy accessibility to family members living afar silently ceases to become the main purpose of owning the phone. A cell phone is used for purposes *otherwise*. To understand these purposes, one must understand the existential predicament arising from the particular background the new migrant workers find themselves in.

As earlier observed, the new migrant workers in Tangxia Town greatly outnumber the local population; yet the two groups of people are almost isolated from each another. The fact is that the workers do not belong to the city (which is in fact only a town), and their monotonous working life on the production line gives them hardly any exposure of city life; rather, it tends to lead to feelings of rootlessness and helplessness. The migrants have a great desire to develop social relationships over which they have some control, and it must be a new grouping from their surroundings — this is the reason why we think they have come to reply on cell phones. Here, we must come back to the meaning of SMS messages to this group of users. Besides being a comparatively cheap way of contacting people, SMS messages also give the users a sense of privacy. The sending and receiving of messages is a silent act, a very different kind of social capturing than what we regard as "civic inattention" as mentioned in the first part of this paper. As one of the interviewees revealed, it is only through the use of SMS that one can lie in bed in the dormitory and *talk* with one's lover when other roommates are present.[14] At the same time, one has a sense of actively controlling technology for one's convenience. This is a very important feeling, giving the user a kind of psychological satisfaction that can counterbalance their low economic status and spartan living environment. The cell phone, simply put, helps to rebuild their self-esteem.

This explains why the sending and receiving of SMS messages via cell phones has become a significant pattern of communication among the new migrant workers.[15] In fact, all except one respondent thus far told us that they would always reply to an SMS message sent by a friend. Of course, important messages from significant others are always treasured. But more often than not, the messages that are sent or received are only greetings, regards or humorous works with interesting animations that help them kill time after

[14]Interview with YWM in the morning session of July 21st, 2003.

[15]Here, the authors have no intention to hint that the rapid expansion in the use of SMS messages by mobile phone users as mentioned earlier is due to the role played by the new migrant workers. Rather, we are attempting, at most, to show how the social meaning given to SMS message by this group of people may be one of the significant features contributing to the expansion in the use of SMS messages.

office hours. It is thus no exaggeration to say that SMS culture is a strong explanation for the high demand on cell phone usage. It is a response to the particular psychological, physical, and socio-economic backdrop in which the new migrant workers find themselves. SMS culture is, after all, a cyber culture; it gives the migrants a new way of interacting to which they have hitherto had no similar exposure; it is something they "actively chose" to be involved in within a socio-economic context that they have little power of manipulation or say. We believe that it is one way they actively construct a *lingxi changyu* (connected sphere) to counteract the "cultural insulation" they are experiencing in a place into which they cannot successfully merge, however much they may aspire to. In this light, the cell phone has become a technological device that can at least help them delay the pain generated from the kind of existential predicament in which they are caught. Alternatively, in a more positive manner, the use of cell phones somewhat bolsters their self-esteem. After all, there is no other consumer durable than a cell phone that can be held in the hand and that will allow them to demonstrate their feelings, confirming their sense of freedom, and assuring their existence as a subject in a rootless time-space.[16]

Rethinking the Interface of Technology and Culture

Until now, we still have not touched on the cultural aspect of the use of cell phones. What we want to argue here is that not only are there specific cultural meanings to the use of cell phones by Chinese in general that can contribute to the study of the sociology of cell phone usage, but that we believe that this kind of study can reveal the importance of meta-theoretical concerns in comparative research around the globe: that is, how far we have paid attention to the notion of universality and particularity in any research question. We will return to this issue in the final section of this paper. First, we try to see what kind of specifications we may detect from the Chinese in general before we put the cultural specifications back into the context of understanding the new migrant workers.

[16]For instance, during the second round of interviews in March 2004, two of our researchers received SMS messages from male informants soon after the interview. The message was about the possibility of setting a date for further meetings. The generation of such an act does not entail a cost for the informants, as this kind of cyber space contact cannot only break through the ice of all class and regional barriers, but also open up chances to interact with "other groupings" that can lead to significant psychological pleasure for a marginal grouping such as theirs.

According to Fei (1992 [English version of his book written in 1947]), the "this-worldly" concern of the Chinese people requires one to think of how to live with other people. To achieve the highest goal of Confucianism, *he* (harmony), one should act in a proper way, and this brings up the idea of *li* (courtesy). Eventually, *lun* (cardinal relationship) became the core element of the Confucian idea, which stresses the importance of *bei* (differentiation). Against this backdrop, the important step for a Chinese is to know how to *tui* (push or extend out), just as ripples flowing out from the center of a splash when a stone is thrown into the water. This is not simply a metaphor, but captures the essence of the Chinese way of interacting in the form of *chaxugeju* (differentiated mode of association): the self, embedded in social relationships and emotionally tied to personal obligations as defined by those relationships, acts out with the required *li* according to that relationship. Here, we see that there is an understanding behind this process: the idea is to treat a person *not* as a person per se (that is, not as an individual in equal terms), but according to the relationship of that person with oneself. This is why the chief characteristic of Chinese society is *lunli benwei* (ethics-based), which means that the focus of the social system of China is on the particular nature of the relationship between those interacting individuals. This makes a normal Chinese, in King's term (1985), a "relational being." Interpersonal relationships must then be measured, and this measuring in turn sets the ethical standard one has to take up in dealing with the other person. As with Fei, Francis Hsu illustrated long ago that Westerners are "individual-centered" whereas the Chinese are "situation-centered" (Hsu 1981 [1953]). This view also has the full support of scholars from the West (Bellah *et al.* 1985, 1992; Madsen 1984; Potter and Potter 1990). This kind of differentiation is not a kind of judgment on which measure is better, but one of showing that there *is* a difference. This is a diversity that has to be acknowledged if we genuinely want to capture the normative values and behavioral norms of the people.

The studies of Fei and Hsu represent academic efforts undertaken more than half a century ago. But for researchers of China, the notion of *renqing* (personal affection), *mianzi* (face), and *guanxi* (social relationships), which are all core cultural elements belonging to the Confucian paradigm, have never been out of the agenda in decades of academic research, whether on carried out in cities or villages, that try to capture the dynamics of Chinese society.[17] The meaning of these cultural traits to this study has to be articulated here.

[17]To name just a few, see Chan *et al.* (1984), Gold (2002), King (1991), Lu and Perry (1997), Madsen (1984), Walder (1986), Yan (1996) and Yang (1994).

It is here that we try to further sharpen our research concern: we want to argue that, to a certain extent, the massive consumption and usage patterns of cell phone by this group of new migrant workers has to be set against a time-space juncture where their cultural traits intertwine with the technology in the particular socio-economic backdrop in which they find themselves situated. As argued above, the new migrant workers generally appear to be very different from their predecessors in terms of the economic, social, and psychological aspects of their lived experiences. To enrich our understanding from a cultural dimension, the massive consumption and use of cell phone by the new migrant workers is also a kind of "social competition" that they carry out among themselves. As one of the factory owners told us, "Boys today will be looked down if they do not have a cell phone in hand. The more advanced the model the more "face" (*mianzi*) you gain;[18] in the past your status was proved by the suit you wear, now it is cell phone that counts. The younger the worker, the greater this desire ... if you are asked for a number and you cannot offer one, you will lose face."[19] The ownership and the pattern of use of a cell phone is itself a symbol of basic achievement, a kind of social position. Yet, this kind of competition for achievement is a pragmatic one: one would only make the comparison with those who have a similar background. The workers know very well that the fact that an entrepreneur may not have a cell phone, or may have an obsolete model, does not affect his/her status as an entrepreneur. So it is meaningful to make the comparison across contemporaries, but it would be silly to do so across "classes!" Seen in this light, it is not for the purpose of communication alone that this group of new migrant workers owns and uses cell phones. The rule of the game is simple: the first level of differentiation is between having a cell phone or not, and the second level of comparison is between owning an updated model or an obsolete one. This explains why a Nokia or Sony Ericsson model is more desirable than one of the national brands, and a color display is preferred to a black-and-white model.

By the same token, we may grasp another kind of cultural meaning for the massive use of SMS messages by this group of new migrant workers. This shed further light on the popular use of SMS messages as mentioned earlier. Sending and responding to SMS messages is not just for the sake of killing time; there is a deeper meaning behind these activities: responding to an SMS message is one way of maintaining a relationship, an act of *bao* (reciprocity) full

[18]It is impossible to go into detail in this discussion on the importance of the notion of "face" to the Chinese. For a classic analysis of the issue, see Hu (1944).
[19]Interview with informant CJ in the morning session on July 22nd, 2003.

of Chinese cultural wisdom.[20] To the workers, even in such a highly developed and highly commercialized area, one can barely survive without an appropriate *guanxiwang* (relationship network).[21] This explains the strong tone behind the statement of "so long as the SMS messages are from friends, we must reply." The closer the relationship between two persons, the more frequent the SMS exchange between them, and vice versa. Resource networks are produced and accumulated via technology, and a newly established "differentiated mode of association" is thus reproduced through an ICT revolution.

Reflections

In this chapter, we have concentrated on studying the consumption and usage patterns of cell phones by an emerging new group of contemporary Chinese — the new migrant workers. We picked this area of research because we observed that the phenomenon of the rapid spread of the use of cell phones in mainland China has not received the attention it should be receiving in the sociological literature. Our research group hopes that our series of discussions may lead to more in-depth analyses on the particular meaning of cell phone usage to the Chinese that take into account the socio-economic and historical-cultural backdrop of the users. In this particular context, we have tried to outline the uniqueness of the new migrant workers, focusing on the particular social time and position in which they find themselves, and how such uniqueness has led them to become consumers of cell phones in a particular manner. Certainly, every individual can use a cell phone for the purpose of communication, but different groupings may make use of ICTs for different ends. There are some universal elements here on which Western sociological studies can shed light, but there are also particular elements that need a specific indigenous understanding. For instance, there is a worldwide trend in the use of "grooming calls" to show concerns, solidarity, and commitment, and to articulate nearness, compassion, sympathy, and love;[22] but in our research field, "grooming calls" appeared in the form of "grooming messages," and there was a great deal of social and cultural meaning attached to both sending and responding to SMS messages. The act captures an existential predicament

[20] For a classic discussion of the concept of *bao*, see Yang (1957).

[21] As mentioned in footnote 17, there have been quite a number of discussions on how *guanxi* influences the behavior of the Chinese. What our research group is focusing on is how this notion is working in a cyber context, or how culture is interfacing with technology in the lives of contemporary Chinese; see Law (2003), Law and Peng (2004), this chapter.

[22] See the discussion in various chapters of Katz and Aakhus (2002).

of this group of people: in a place where they have been marginalized and into which they have little hope of becoming anchored in the long run, to get a taste of manipulating technology at one's discretion is an extremely meaningful experience. In this way, a gloomy future is balanced by the psychological pleasure gained from controlling a modern technological product. It is in this particular socio-historical juncture that we find the explanation to such a pattern of consumption pattern in a town in south China. After all, such an engagement in ICTs is the closest they have gotten to what is meant by an active agent in cyber space, a symbol of city life, a breath of individual style. Yet the cell phone as a product of late modernity draws them magnificently back to fundamental cultural traits that are too costly for them to ignore. The technology surrounding the use of the cell phone has provided them with the necessary convenience to make use of it for maintaining relationships, developing networks, and strengthening bonds. In this way, the normative values and behavioral norms of traditional Chinese culture have entered cyber space, although how it is going to evolve is yet to be seen.

There is one last word on the kind of methodological concern that we can derive from the discussion. As the use of the cell phone is a global phenomenon, is there a universal understanding of how to approach the threads behind this human activity? More than a decade ago, in the presidential address to the 12th World Congress of Sociology, Archer advocated a single sociology for one world (Archer 1991). Those who read Archer even in brief could clearly sense her strong urge to deal with the irony: an increasingly global society is being met by an increasingly localized sociology. Perhaps this is due to the two-faced nature of globalization. On the one hand, the closer we are brought to one another in the world, the more we know that, to borrow Cohen's phrase, the equation "modern is Western, and Western is important" does not work (Cohen 1984). On the other hand, unless one is denying the process of globalization, one has to admit that this process, as constitutive of social reality everywhere, is the very aspect that cannot be understood in strictly local terms itself. In this research paper, we try to spell out that there is something particular about the use of cell phones by Chinese that needs to be explored further. But does this mean that we are playing tribute to indigenous awareness, and thus declaiming Western sociological applications in the Chinese context? Or is globalization a phenomenon that is giving us a new meaning of modernity that takes a more holistic approach (what Archer tags as "theoretical variety") to understanding our world; or, on the contrary, is it telling us that besides the West there is still the rest, so the very fact of going global is ironically paving the way for celebrating the local? We do not want to be drawn into the fierce debate over this subject in the current sociological literature, but the fact that we are raising concerns relating to the indigenous is that we believe that the indigenization of sociology is for the benefit of the whole field. Simply put, all social science researchers are trying to understand the reasons behind certain social phenomena. Yet, all of these studies must take the reflections of the concerned researcher one is situating in as a starting point. In other words, the moment we adopt a research methodology or a theoretical framework, its use implies that we have accepted the presuppositions of that framework on the human self and society. If we look deep into the structures of the social science theories of the West, to some extent they have taken certain foundational aspects of human life for granted as a starting point. This is a product of their culture, their philosophical traditions and theological concerns. The basic understanding of human beings, and the kind of value systems derived from such an understanding, is the consequence of a kind of evolution based on the cultural traditions and societal changes experienced by a particular society.

As shown in the discussion here, in the course of examining how the Chinese use mobile telephony, we must understand that what is behind the consumption and usage patterns is embedded with deep cultural meanings. Once this is understood, this inevitably touches on how Chinese handle relationships with others. Do the Chinese behave like Westerners then? If we, as indigenous researchers, cannot take note of the different perceptions of self of the Chinese,[23] then perhaps we are simply wasting the chance to improve the explanatory power of sociology for all by reviewing the applicability of sociological knowledge in various indigenous communities. But, of course, we are not saying that only Chinese can understand Chinese. Fei Xiaotong and Francis Hsu could compare Americans with the Chinese half a century ago because they did not think of the Americans in terms of their own culture, and accept that there is another set of underlying values and assumptions in which Americans do believe and by which reflect upon their own day-to-day activities. At the end of the day, the first principle for an authentic comprehension in all kinds of cross-cultural studies is perhaps an empathetic understanding of the researchers in a genuine sense.

[23]See the brilliant discussion on this subject by Liu (2002).

References

Archer, M. S. 1991. "Sociology for One World: Unity and Diversity." *International Sociology* 6(2):131–147.

Aronson, S. 1971. "The Sociology of the Telephone." *International Journal of Comparative Sociology*, 12:153–167.

Beck, U. 1992. *Risk Society: Towards a New Modernity.* London: Newbury Park.

Bellah, R. N. *et al.* 1985. *Habits of the Heart: Individualism and Commitment in American Life.* New York: Harper and Row.

Bellah, R. N. *et al.* 1992. *The Good Society.* New York: Vintage Books.

Chan, A., R. Madsen and J. Unger. 1984. *Chen Village: Under Mao and Deng.* Berkeley: University of California Press.

Cohen, P. A. 1984. *Discovering History in China: American Historical Writing on the Recent Chinese Past.* New York: Columbia University Press.

Cooper, G. 2000. "The Mutable Mobile: Social Theory in the Wireless World." Paper Presented at the "Wireless World" Workshop, April 7th, University of Surrey.

De Sola Pool, I., ed. 1977. *The Social Impact of the Telephone.* Cambridge, MA: MIT Press.

Dimmick, J., J. Sikand and S. Patterson, 1994. "The Gratifications of the Household Telephone: Sociability, Instrumentality and Reassurance." *Communication Research*, 21(5):643–663.

Fei, X. 1992. *From the Soil: The Foundations of Chinese Society.* California: University of California Press.

Fortunati, L. 2002. "Italy: Stereotypes, True and False." In *Perpetual Contact: Mobile Communication, Private Talk, Public Performance*, edited by J. E. Katz and M. Aakhus. Cambridge: Cambridge University Press.

Fox, K. 2001. "Evolution, Alienation and Gossip: The Role of Mobile Telecommunications in the 21st Century." Social Issues Research Center, Oxford. http://www.sirc.org/publik/gossip.shtml

Castells, M., ed. 2004. *The Network Society: A Cross-cultural Perspective.* Cheltenham: Edward Elgar.

Castells, M. 2000. *Information Age Volume One: The Rise of the Network Society*, 2nd Ed. Oxford: Blackwell.

Geser, H. 2004. *Towards a Sociological Theory of the Mobile Phone.* Switzerland: University of Zurich, Institute of Sociology. http://socio. ch/mobile/t_geser1.htm

Giddens, A. 1990. *The Consequences of Modernity.* Cambridge: Polity Press.

Gold, T. *et al.*, eds. 2002. *Social Connections in China: Institutions, Culture and the Changing Nature of Guanxi.* Cambridge: Cambridge University Press.

Green, N. 2003. "Community Redefined: Privacy and Accountability." In *Mobile Communication: Essays on Cognition and Community*, edited by K. Nyiri. Vienna: Passagen Verlag.

Hu, H.-C. 1944. "The Chinese Concept of 'Face'." *American Anthropologist* 46(1):45–64.

Hsu, F. L. K. 1981[1953]. *Americans & Chinese: Passage to Differences*, 3rd Ed. Honolulu: University of Hawaii Press.

King, Y.-C. 1991. "The Individual and Group in Confucianism: A Relational Perspective." In *Individualism and Holism: Studies in Confucian and Taoist Values*, edited by D. Munro. Ann Arbor: Center for Chinese Studies, University of Michigan.

King, Y.-C. and M. H. Bond. 1985. "The Confucian Paradigm of Man: A Sociological View." In *Chinese Culture and Mental Health*, edited by W.-S. Tseng and D. Wu. Orlando: Academic Press.

Kopomaa, T. 2000. *The City in Your Pocket — Birth of the Mobile Information Society*. Helsinki: University Press Finland.

Lasen, A. 2002. "The Social Shaping of Fixed and Mobile Networks: A Historical Comparison." Digital World Research Centre, University of Surrey.

Law, P.-L. 2003. "Research Notes on Cellphones and the Social Lives of Migrant Workers in southern China." Presented in the Conference of Cultural, Social and Political Consequences of the Computer Mediated Interactive Communication Technologies, October 20th–22th, Manila, The Philippines.

Law, P.-L. and Y. Peng. 2004. "Cell Phone, Internet and the SARS Epidemic." Paper Presented in the Session the Co-construction of Society and Technology in a Global World. International Institute of Sociology 36th Congress, July 7th–11th, Beijing.

Liu, X. 2002. *The Otherness of Self: A Genealogy of the Self in Contemporary China*. Ann Arbor: University of Michigan Press.

Lu, X. and E. J. Perry, eds. 1997. *Danwei: The Changing Chinese Workplace in Historical and Comparative Perspective*. Armonk: M.E. Sharpe.

Luo X., and C.-G. Wang. 2003. "Xinshengdai Nongcun Liudong Renkou de Waichu Dongyin yu Xingdong Xuanze." (The Working Out Motives and Choice of Action of the New Generation of Village Migrants), *Zhejiang Shehui Kexue*, No. 1.

Lyon, D. 1988. *The Information Society: Issues and Illusions*. Cambridge: Polity Press.

May, C. 2002. *The Information Society: A Sceptical View*. Cambridge: Polity Press.

Madsen, R. 1984. *Morality and Power in a Chinese Village*. Berkeley: University of California Press.

Puro, J.-P. 2002. "Finland: A Mobile Culture." In *Perpetual Contact: Mobile Communication, Private Talk, Public Performance*. Cambridge: Cambridge University Press

Qui, J. L. 2004. "The Internet in China: Technologies of Freedom in a Statist Society." In *The Network Society: A Cross-cultural Perspective*, edited by M. Castells. Cheltenham, Northampton: Edward Elgar.

Ross, J. P. 1993. "300000 yuppies? Mobile Telephones in Finland." *Telecommunications Policy* 17(6):446–458.

Townsend, A. M. 2000. "Life in the Real-time City: Mobile Telephones and Urban Metabolism." *Journal of Urban Technology* 7(2):85–104. http://www.informationcity.org/research/real-time-city/life-in-the-real-time-city.pdf

Walder, A. G. 1986. *Communist Neo-tradition: Work and Authority in Chinese Industry*. Berkeley: University of California Press.

Wang, C.-G. 2001. "Xinshengdai Nongcun Liudong Renkou de Shehui Rentong yu Chengxiang Ronghe de Guanxi." (The Social Identification of the New Generation of Village Migrants and Its Relationship with City-village Mergers), *Shehuixue Janjiu* 3:63–76.

Webster, F., ed. 2004. *The Information Society Reader*. London and New York: Routledge.

Webster, F. 2002. *Theories of the Information Society*, 2nd Ed. London and New York: Routledge.

Wellman, B. 2001. "Physical Place and Cyber Place: The Rise of Personalized Networking." *International Journal of Urban and Regional Research* 25:227–252.

Yan, Y. 1996. *The Flow of Gifts: Reciprocity and Social Networks in a Chinese Village*. Stanford: Stanford University Press.

Yang, L.-S. 1957. "The Concept of Pao as a Basis for Social Relations in China." In *Chinese Thought and Institutions*, edited by J. K. Fairbank. Chicago: University of Chicago Press.

Yang, M.-H. M. 1994. *Gifts, Favors and Banquets: The Art of Social Relationships in China*. Ithaca and London: Cornell University Press.

Yu, L. A. and T.-H. Tng. 2003. "Culture and Design for Mobile Phones for China." In *Machines that Becomes Us: The Social Context of Personal Communication Technology*, edited by J. E. Katz. New Brunswick, New Jersey: Transactions Publishers.

Appendix I. Number of Phone Users by Province

Source: Monthly statistics as provided by MII

* The figures in brackets are data announced in July 2003; otherwise they represent figures in December 2004
* The highlighted percentages represent the percentage increase/ decrease in of the number of phone users in that category between July 2003 and December 2004
* For the July 2003 data, see http://www.mii.gov.cn/mii/hyzw/tongji/yb/tongjiyuebao20030701.htm
* For the December 2004 data, see http://www.mii.gov.cn/mii/hyzw/tongji/yb/tongjiyuebao200412.htm

Unit: 10,000 users

Province	Fixed Phone			Cell Phone
	Total	City Phone	Village Phone	
Nationwide	**(24075.4)** **31244.3** **29.78%**	**(15328.3)** **21084.8** **37.55%**	**(8747.1)** **10159.5** **16.15%**	**(23945.9)** **33482.4** **39.83%**
Eastern Area	(11591) 15041.0 29.76%	(7353.9) 10348.2 40.72%	(4237.1) 4692.8 10.75%	(12870.1) 17411.7 35.29%
Beijing	(620.7) 847.4 36.52%	(540.9) 759.7 40.45%	(79.7) 87.7 10.04%	(1032.5) 1340.7 29.85%
Tianjin	(314.2) 407.2 29.60%	(313.2) 405.9 29.60%	(1) 1.3 30.0%	(359.7) 423.6 17.76%
Liaoning	(1128.2) 1472.3 30.50%	(782.2) 1060.9 35.63%	(346) 411.5 18.93%	(998.3) 1194.4 19.64%
Shanghai	(701.9) 867.5 23.59%	(701.9) 858.5 22.31%	(0) 9.0 900%	(1018.8) 1311.3 28.71%
Jiangsu	(1875.9) 2582.4 37.66%	(1027.7) 1641.9 59.76%	(848.2) 940.5 10.88%	(1801.6) 2232.9 23.94%
Zhejiang	(1543.5) 1974.8 27.94%	(850.9) 1188.2 39.64%	(692.5) 786.6 13.59%	(1701) 2322.5 36.54%
Fujian	(1070.8) 1265.8 18.21%	(649) 840.6 29.52%	(421.8) 425.2 0.88%	(839.5) 1138.1 35.57%
Shandong	(1956) 2464.1 25.98%	(908.6) 1265.8 39.31%	(1047.4) 1198.2 14.40%	(1347.9) 1909.4 41.66%

(*Continued*)

(*Continued*)

Province	Fixed Phone			Cell Phone
	Total	City Phone	Village Phone	
Guangdong	(2220.2) 2962.3 33.42%	(1466.3) 2186.2 49.10%	(753.9) 776.2 2.96%	(3645) 5373.9 47.43%
Hainan	(159.7) 197.3 23.54%	(113.2) 140.6 24.20%	(46.5) 56.6 21.72%	(125.5) 165.0 31.47%
Central Area	(7708.3) 9745.7 26.43%	(4602.9) 6074.1 31.96%	(3105.3) 3671.6 18.24%	(6417.6) 9150.7 42.59%
Hebei	(1257) 1554.9 23.70%	(668.8) 883.1 32.04%	(588.2) 671.8 14.21%	(1060.9) 1512.9 42.61%
Shanxi	(625.3) 770.9 23.28%	(425.3) 521.9 22.71%	(200) 248.9 24.45%	(504.4) 753.8 49.44%
Jilin	(544.7) 655.6 20.36%	(384.9) 461.5 19.90%	(159.9) 194.1 21.39%	(573.9) 763.8 33.09%
Heilongjiang	(793.6) 1082.6 36.42%	(591.3) 833.0 40.88%	(202.4) 249.6 23.32%	(764.4) 1017.1 33.06%
Anhui	(917.5) 1191.6 29.87%	(423) 591.9 39.93%	(494.4) 599.7 21.30%	(640.2) 873.4 36.43%
Jiangxi	(580.5) 747.4 28.75%	(349.3) 461.2 32.04%	(231.2) 286.1 23.75%	(490.1) 671.3 36.97%
Henan	(1274.1) 1583.8 24.31%	(690.5) 927.2 34.28%	(583.6) 656.7 12.53%	(923.4) 1392.3 50.78%

(*Continued*)

(*Continued*)

Province	Fixed Phone			Cell Phone
	Total	City Phone	Village Phone	
Hubei	(823.6) 1073.6 30.35%	(587.9) 777.8 32.30%	(235.7) 295.8 25.50%	(712.8) 1129.8 58.50%
Hunan	(891.9) 1085.3 21.68%	(481.9) 616.5 27.93%	(410) 468.8 14.34%	(747.5) 1036.4 38.65%
Western Area	(4774.1) 6267.5 31.28%	(3369.7) 4491.8 33.30%	(1404.4) 1775.6 26.43%	(4658.2) 6920.0 48.56%
Inner Mongolia	(372.6) 490.7 31.70%	(293.6) 388.4 32.29%	(79) 102.3 29.49%	(418.9) 594.6 41.94%
Guangxi	(596.8) 812.3 36.12%	(406.9) 545.7 34.11%	(189.9) 266.6 40.39%	(480.5) 874.5 82.00%
Chongqing	(469.1) 626.4 33.53%	(294.7) 425.5 44.38%	(174.4) 216.9 24.37%	(502.6) 811.6 61.48%
Sichuan	(1013.6) 1369.9 35.15%	(710.4) 989.2 39.25%	(303.2) 380.7 25.56%	(1017.5) 1514.6 48.86%
Guizhou	(312.7) 388.6 24.27%	(223.3) 282.0 26.29%	(89.5) 106.6 19.11%	(313.7) 440.0 40.26%
Yunnan	(469.6) 547.4 16.63%	(348.4) 414.8 19.05%	(121.2) 132.6 9.41%	(534.5) 732.4 37.03%
Tibet	(23.7) 37.7 59.07%	(22.4) 35.7 59.38%	(1.3) 2.0 52.85%	(29.1) 39.7 36.43%

(*Continued*)

(*Continued*)

Province	Fixed Phone			Cell Phone
	Total	City Phone	Village Phone	
Shaanxi	(614.5) 792.0 28.88%	(403.9) 527.2 30.53%	(210.6) 264.8 25.74%	(540.4) 788.7 45.95%
Gansu	(367.6) 477.3 29.84%	(259.5) 344.6 32.79%	(108.1) 132.7 22.76%	(261.8) 358.0 36.75%
Qinghai	(68.5) 94.4 37.81%	(58.2) 77.7 33.51%	(10.3) 16.7 62.12%	(92.2) 117.7 27.66%
Ningxia	(92.4) 119.6 29.44%	(68.3) 84.4 23.57%	(24.1) 35.2 46.06%	(105.4) 158.6 50.47%
Xinjiang	(373) 495.2 32.76%	(280.2) 376.7 34.44%	(92.9) 118.6 27.66%	(361.6) 489.7 35.43%

Appendix II. Figures Concerning the Use of SMS Messages by Mobile Phone Users

Source: Statistics for the months concerned as provided by MII, see http://www.mii.gov.cn

Period	SMS messages (unit: 100,000,000)	Total number of mobile phone users during the period (unit: 10,000)	Average SMS messages sent out by each mobile phone during the period
01/2004	156.6	27680.2	56.57
02/2004	317.3	28232.7	112.38
03/2004	483.7	29030.5	166.62
04/2004	648.2	29575.0	219.17
05/2004	822.0	30055.9	273.49
06/2004	996.3	30528.3	326.35
07/2004	1178.5	31021.8	379.89
08/2004	1363.5	31510.0	432.72
09/2004	1558.5	32007.1	486.92
10/2004	1760.6	32503.4	541.66
11/2004	1958.9	32992.4	593.74
12/2004	2177.6	33482.4	650.37

Appendix III. Comparison of Users and Penetration Rate of Fixed Phones and Mobile Phones

Source: Statistics for the months concerned as provided by MII, see http://www.mii.gov.cn

Period	Total number of fixed phone users during the period (unit: 10,000)	Penetration rate of fixed phone during the period (unit: no. of phones owned per 100 persons)	Total number of mobile phone users during the period (unit: 10,000)	Penetration rate of mobile phones during the period (unit: no. of phones owned per 100 persons)
01/2003	21800.4	17.5	21243.9	16.2
04/2003	22903.9	17.5	22571.7	16.2
07/2003	24075.4	19.4	23945.9	18.3
10/2003	25513.9	20.3	25693.8	19.5
01/2004	26893.3	21.2	27680.2	20.9
04/2004	28544.8	21.2	29575	20.9
07/2004	29899.6	23.7	31021.8	23.7
10/2004	31069.5	24.5	32503.4	24.8
12/2004	31244.3	24.9	33482.4	25.9

11 *The Use of Mobile Phones among Migrant Workers in Southern China*

Pui-lam Law and Yinni Peng

Introduction

Migrant workers are among the lowest-income groups in Guangdong, taking on largely skilled and semi-skilled jobs in factories with monthly salaries ranging from 400 to 800 yuan (US$1 is equivalent to approximately 8.23 yuan). Despite the fact that most migrant workers belong to the lowest stratum in Guangdong, the number owning mobile phones has increased rapidly since 2002. These migrant workers use mobile phones largely to keep in closer contact with their families at home, to maintain their networks with relatives and friends working in Guangdong, and to make new connections in cyberspace. In view of these phenomena, the aim of this paper is to explore the social consequences of mobile phone use among migrant workers in southern China. This paper is developed mainly as the result of a study into the social impact of mobile phone use among migrant workers in industrialized villages in Dongguan City, Guangdong Province.[1] The study has been on-going since July 2003 and the migrant workers interviewed are young villagers from less-developed or poor provinces such as Hunan, Sichuan, and Guangxi.

Since mid-2003 onwards, we have conducted interviews with migrant workers in Dongguan on their use of mobile phones. We chose workers in Dongguan for the following reasons. According to data from the Ministry of Information Industry (2004), mobile phone penetration rates in Guangdong Province are the highest in the country. Among the cities in Guangdong,

[1]This study was funded by a grant from The Hong Kong Polytechnic University. We would like to thank Liu Xiaojing, for arranging the interviews in Tangxia and Humen Townships and for his assistance during the interviews. Thanks also go to the interviewees as, without them, this study would not have been possible.

Dongguan City is considered to be a city of migrants, as migrant workers make up over 5 million of the total population of 7 million. These workers come from different provinces and are working in various kinds of factories. In other words, Dongguan can serve as a typical city in studying mobile telephony and migrant workers in southern China. During our study, we had thought of conducting a survey. However, to have a reliable sample size representative enough to generalize theoretically about the whole population, a considerable amount of resources, including both money and manpower, would have been required. In addition, due to the highly vagrant nature of the workers, a sample household survey would have been almost impossible to conduct. To conduct structured interviews in a factory, we would have had to have a very good relationship with the proprietors of the factory. Even if it had been possible to develop such rapport, the resulting data might not have been that reliable, as the workers would probably have believed that the interviews were being conducted on behalf of their proprietors and they might therefore not have been willing to provide reliable information. We therefore had to rely on in-depth interviews and participant observations. By the end of 2004, we interviewed over 40 migrant workers. The workers came from different factories in the villages of Dongguan City. At least 28 workers were interviewed three times since 2003. Of these migrant workers, 14 were male and 14 female, ranging in age from 16 to 30. Although the size of the sample is small, the interviews inform some essential themes that are central to the study of the social consequences of the adoption of mobile phones among migrant workers in the southern part of China.

Before discussing the social consequences of the adoption of mobile phones among the migrant workers, a few words about the theoretical discussion of this new communication technology may be useful. The Katz and Aakhus compendium (2002) points out that the mobile phone will significantly change people's lives and their social relationships both directly and indirectly, since this revolutionary technology is a personalized mobile device (Green 2003) that frees the individual from physical and temporal constraints (Kopomaa 2000) and provides both oral and written (short message service) modalities of communication (Pertierra *et al.* 2003). With the features of instant and perpetual connectivity, the mobile phone can continuously link individuals who are physically absent, hence forming an absent-presence cyber network (Gergen 2003). Thus, Pertierra *et al.* (2003) argued that mobile phones maintain existing social relationships in expanded spatio-temporal contexts. In light of these theoretical ideas, this paper discusses, in the following order, how the adoption of mobile phones among migrant workers will change both their social lives and their social relationships. The first section will describe the mobile phone consumption patterns of migrant workers, showing how the mobile phone is changing the social lives of the migrant workers. The second section will discuss how workers are maintaining their existing social relationships and developing new ones in an expanded spatio-temporal context. The final section is the conclusion.

Mobile Phone Consumption Patterns

The growing popularity of mobile phones among migrant workers has had a huge impact on the patterns of mobile phone consumption. During the first stage of fieldwork in July 2003, we found that migrant workers usually purchased new models of mobile phones, if not the latest, and mostly well-known imported brands such as Nokia, Motorola, and Samsung. The price of the kinds of mobile phones they bought averaged between 1,500 and 2,000 yuan. They bought such phones even though they earned between 400 to 800 yuan a month and despite the fact that many good models costing less than 800 yuan would certainly have sufficed to help them keep in touch with their families and friends. However, the design, look, function and, more importantly, the brand name seemed to be primary considerations when the workers chose their mobile phones. Thus, when we carried out interviews of migrant workers in early 2004, most had mobile phones with colour display, and some even had mobile phones with a camera function even though they did not find it very useful. The price of these kinds of mobile phones was over 2,000 yuan, with some even being over 3,000 yuan; yet their monthly incomes have not increased much since 2003. Thus, when workers selected mobile phones, they based their choice on price, functions, and brand name.

There are three types of mobile phone on the market. The first consists of imported brands such as Motorola; the second is where the case is made in China and the chips are imported mainly from South Korea; and the third is where both the case and the chips are made in China. In 2003, nearly all of the workers interviewed had purchased imported brands, while in 2004 we found that more workers had purchased the second type of mobile phone because they were cheaper, the functions were better, and the parts were as reliable as imported brands. For instance, a mobile phone of the second type costing 2,300 yuan would have a camera with a resolution of 33k pixels while imported models selling at the same price would have a camera function with a resolution of only 16k pixels. Although most of the workers would have liked to have an imported brand, if they do not yet have enough money to purchase such a phone and want to have better functions, then they have to compromise and purchase the second type of mobile phone.

In recent years, with improvements in technology and severe market competition, mobile phone models have changed rapidly. Buying a mobile phone usually costs three to four times a migrant workers' salary; therefore, they choose their mobile phones very carefully so that their mobile phones have new functions and will not become outdated quickly. Yet however carefully the workers choose their mobile phones, in only three or four months new models with better designs and functions come on the market, and are sometimes even cheaper than the models they have purchased. Usually, the workers who have bought mobile phones in the previous three or four months become very unhappy with their choice because their mobile phones are already out of date. Some who cannot resist the temptation to have the latest model to show off will sell their old model on the second-hand market and get themselves the most fashionable one.

Workers are also fond of choosing phone numbers. The normal storage value of a phone card is 100 yuan, but a phone card will sometimes cost more than this if it contains a good phone number. A phone number is considered good mainly if it contains digits that are considered lucky or are easy to memorize. For instance, a phone card with the number "13712976013" is difficult to memorize so it costs just 100 yuan. The number "13712476014" also costs 100 yuan because it contains the unlucky digit "4", which means "death" because the pronunciation of "4" is similar to the pronunciation of "death" in both Cantonese and Mandarin. Chinese, particularly the Cantonese, love the digit "8" because it represents getting rich. The combination of the digits 2 and 8 is lucky because it means getting rich easily. 13668 is also a very good combination because it says that you will be rich wherever you go. So "13668282828" is a very lucky number and will cost more than a thousand yuan. We found one worker who had spent 800 yuan on a phone card with a good phone number.

Nowadays, most factories have installed landline phones in their dormitories for the migrant workers. Some have several phones on each floor and some even have one in each room. These are more easily accessible than before, yet workers seem to prefer using mobile phones. The average monthly expense for a worker using a landline is around 20 to 30 yuan, but the average monthly expense for a worker using a mobile phone is around 100 yuan and some obsessed users even spend up to 200 yuan.

In 2003, both of the two major national network providers, China Mobile and China Unicom, charged 60 cents per minute for making or receiving calls. For sending messages, China Travel, part of China Mobile, charged the sender 15 cents a message while China Unicom only charged 10 cents. China Mobile had been promoting a new storage card called M-Zone, which costs 20 yuan a month to send 300 messages. Due to severe market competition, these networks dropped their rates to 39 cents per minute for both local and distant calls from June 2003. Since 2004, China Mobile has charged even less. It charges 39 cents during the day and 19 cents after midnight for making or receiving calls within Guangdong Province. The M-Zone network charges 15 cents during the day and 10 cents after midnight for making and receiving calls on the M-Zone network but 30 cents during daytime and 15 cents after midnight if making calls to or receiving calls from other networks. The regional networks charge even less. The People regional network requires users to pay only 26 yuan per month, which covers all calls received; and charges 18 cents during the day and 9 cents after midnight for making calls.

However, the downward trend in charges has not caused the workers to consume less. A worker who subscribes to a regional network still spends up to 100 yuan a month. It is clear then that workers have become so used to the perpetual contact that mobile phones provide that cheap rates have only caused them to become more frequent users instead. In fact in 2003, when we interviewed the workers we found that they were very cautious about using their mobile phones to contact their families. When they received a call, they made a point of first finding out who was calling by using the caller display function, before answering the call. If the call was from home, by prior arrangement they would not answer, but would go to a public phone booth call their families back. Only in an emergency would they answer a call from home using their mobile phone. This kind of prudent arrangement is no longer common among the workers. It seems that the convenience provided by the mobile phone has turned them into frequent users. Thus, some workers are spending more even though the rate has dropped.

As mentioned above, in addition to two major national network providers, there are regional networks in Guangdong Province, which are cheaper than the two major providers. Yet cheaper rates do not necessarily lead to higher subscription rates as different kinds of migrant workers subscribe to different kinds of networks. Workers that have a more stable job subscribe to the regional networks while workers in less stable jobs subscribe to the China Travel network because the regional networks only provide a service within Dongguan City. Those who are fond of on-line services and of sending messages subscribe to M-Zone.

Nowadays, according to our interviews, many workers lead frugal lives largely because they desire to purchase a mobile phone and to be able to pay the ensuing expenses such as phone services expenses and the cost of more frequent social gatherings arising from mobile phone contact. In recent years as mobile phones have become more popular, more and more workers have become overwhelmed by the desire to own a mobile phone as soon as they arrive in Guangdong. Even though having a mobile phone can help workers to get better jobs, it is unlikely they will see a large rise in their income as most are in semi-skilled jobs. On the other hand, their mobile phone consumption patterns will obviously have an adverse effect on any prudent plans that they may have made. Nowadays, workers have to work for more than ten years to build a new house in their home villages and some cannot even save any money for their family (Law 2003). The fetish for mobile communication among migrant workers has caused them to contribute a large part of their salaries to both the mobile phone manufacturers and network providers.

Mobile Phones and Social Relationships

Mobile phones help maintain existing social relationships in expanded spatio-temporal contexts (Pertierra *et al.* 2002). Kinship is fundamental to the fabric of Chinese social relationships, particularly in rural societies. Kinsmen are defined as "insiders" upon whom one can lay one's trust; others are "outsiders" and have to be measured in instrumental terms (Metzger 1998). When rural workers work in Guangdong, they have to try their best to contact their kinsmen there, as they believe that their kinsmen are the only people upon whom they can rely. Thus, migrant workers in Guangdong will try their best to hold regular social gatherings with their relatives, so as to maintain their connections. The mobile phone, however, pulls together kinsmen scattered far and wide in Guangdong. Through connections in cyberspace, it is more convenient for them to arrange gatherings. Now, they can meet whenever and wherever they wish just by calling each other's mobile phones.

Even though migrant workers cannot meet regularly, they can maintain perpetual contact by using mobile phones. As Gergen has maintained, the mobile phone also serves the function of pulling people who are physically far away into an immediate cyber presence (Gergen 2003). For migrant workers who are "floating" far away from their home villages, mobile phones enhance their ties with family members back in their home villages. For instance, a young female worker said that she could maintain contact with her family even at midnight when she felt lonely. Thus, a form of absent presence can be developed using the mobile phone, which allows people to maintain closer contacts with family members who are physically far apart, forming a kind of free-floating kinship network.

Mobile phones are not only helping migrants to maintain existing kinship relationships in expanded spatio-temporal contexts, but also to prolong new social relationships developed in the workplace. Workers come to know each other when working on the same assembly lines or staying in the same dormitories. They develop friendships among workers from different villages, towns, or provinces even though they are not kinsmen and would normally consider each other to be "outsiders." Thus, industrialization in Guangdong provides a platform for developing a new kind of social relationship, where people are not simply polarized into either "insiders" or "outsiders." Migrant workers are, however, highly mobile and most do not stay in one factory for a long period of time. Once they find other factories offering higher incomes and better welfare provisions, they will leave for these factories immediately. Hence, without mobile phones to make the physically absent present, friendships cannot be prolonged. The growing popularity of mobile phones among migrant workers serves the function of preserving this new kind of social relationship.

The mobile phone is also useful for young workers for the purpose of courtship. In the past, marriages were arranged by parents. After 1949, despite the fact that the government had promulgated new marriage laws and encouraged the freedom to choose marriage partners, in rural areas males and females remained very separate, not daring to speak to strangers of the opposite sex (Parish and Whyte 1978).[2] As Potter and Potter maintained, "Chinese emotions ... are natural phenomena without important symbolic significance for the maintenance of perpetual social relationships" (Potter and Potter 1990). Emotions are sometimes devalued or carefully concealed. Thus, a father should conceal his love for his son and couples should hide their emotional intimacy from each other. Since the 1990s young Chinese have become more expressive than before, but in general migrant workers are from less-developed rural areas and most of them still stick to traditional practices. When a boy is attracted to a girl and wants to express his emotions to her, he would normally prefer writing to speaking face-to-face. In this case, the mobile phone serves as a very important channel of communication. When a male migrant worker intends to court a female worker, he simply sends messages to express his feeling towards

[2]In rural areas, girls and boys were often segregated, and it was a traditional custom that boys and girls should not talk to each other individually. See Parish and Whyte (1978).

the girl or to indicate his desire to date her. Therefore, with the use of mobile phones young migrant workers are now more ready to express their emotions.

Mobile phones also provide an on-line service for people to meet new people in cyberspace. Workers are always receiving messages from strangers hoping to make contact with others through mobile phones. In addition to these messages, they also receive messages from organizations that have been set up by the network providers. One popular network, M-Zone, provides users with an on-line service called Monternet, which offers its subscribers a number of chat rooms for meeting others on-line. These chat rooms, for example, "Dating Lovers," "Atypical Boys and Girls," "X-City Romance," and "Absolute Private Intimate Encounter," are designed to help men and women find romance. Some join this kind of network merely to develop a virtual romance, but some hope to make face-to-face contact. The following are some conversations that have taken place in the chat rooms.

(The name of A is "Lonely wolf from the north" [male]; and B is "Softly swinging flight" [female]).

A: How are you?
B: Fine!
A: After work? Chat for a while?
B: How are you? My name is Ah Yung.
B: Not yet after work, and you?
A: It is my day off today. Ha! Ha! Where do you work?
B: I work in Weizhou.
B: I am from Hubei, where do you work?
A: Have you slept yet? I work in a factory producing golf equipment. How about you? Some kind of predestination is leading us together, you are from Hubei and I am from Hunan.
B: I work in a factory producing glass. I have to do overtime work every night. How about you? Do you need to do any overtime work?
A: I always have overtime work, but I do not need to work tonight. The latest is to work until midnight. When do you go to bed? I have just joined the chat room for a short while and am not very used to it. Ha! Ha! Your name is very special.
B: I am not very familiar with it, too. I have never ever played that before. Usually, it is others who send me messages.
A: How long have you been in Dongguan?
B: Not so long, about a year.
A: How old are you?
B: 19, how about you?

A: Me, 25, single :) Ah Yung, usually what do you do when you are free? I go to karaoke.
B: I stay at home, or hang around the streets.
A: Buying clothes? Girls like to go shopping and buy clothes. Usually where do you go? I am very sleepy and have to go to bed. Can you tell me your mobile phone number so that I can call you if I am free?
B: I don't like buying clothes; I hate buying clothes.
A: You are different from other girls. Then what do you like to do?
B: Okay, go to sleep. Give me your phone number.
B: I love smiling, usually I like hanging around the malls or night markets and buy some stuff, some lovely stuff, or delicious food.
A: You are lovely, I love eating, I like hot spicy food. Okay, my number is: xxxxxxxxxxx.
B: Me too, but not too hot.
A: Sometimes, I miss my family, and you?
B: Yes, I will go back home this year.
A: Usually when do you go to bed? Day off tomorrow? Not disturbing you going to bed?
B: About 2 in the morning, day off tomorrow. And you?
A: Usually 1 o'clock. We have to switch off the lights at 11 o'clock and then lie in bed and send SMSs, and you?
B: Me too, I'm used to it, can't sleep until 2.
A: Do people in your room chat with each other?
B: Sometimes we do, sometimes we don't.
A: We don't, we send SMSs.
B: Do you? We are different, the light won't be switched off during the night and we can read books at night.
A: You read books, what kind of books do you read?
B: Humorous books, books on learning Japanese, and some others.
A: You learn Japanese! Great!
B: No, just because I am bored and having a look only.
(They started exchanging messages at around 9 o'clock at night and they chatted until a quarter past 1 in the morning.)

During the first stage of research conducted in July 2003, most interviewees claimed that they would neither reply to this kind of message nor join the network, but said they knew of other workers who would. When we went back to the villages in 2004, we heard a number of stories of friendships being formed through the mobile phone network. One female worker told us that sometimes she would log on the network in order to make friends on cyberspace. She disguised herself as a man and called herself "Wild wolf." Eventually, a lot of netizens who said they were female, chatted with her and showed an interest in dating her. This female worker found this way of making contact with people interesting but she refused to go out with any of them because she said that this kind of friendship was not serious. To her chatting with the netizens was just for fun and excitement.

Another female worker told us that she enjoyed logging on to the chat rooms and chatting with male netizens although she had a steady boyfriend. The chat rooms she logged on to were mainly designed for workers, for example, "Romance in the Workplace." She said that she used different names, such as "Apple," "Ice," and "Windie," when she logged on to different chat rooms. Usually, she would chat with the netizens for a while and if she found them interesting, they would exchange mobile phone numbers and keep in contact by sending messages. Her record for sending messages was 160 in one day to at least 7 different netizens. She knew that the sweet words and honeyed phrases were an illusion, yet she enjoyed chatting as she found their words comforting. Although she had found that relationships developed through chat rooms were not reliable, she had once dated a netizen working in the same village. She had, however, found it hard to develop their relationship because the mystical feeling achieved in cyberspace had disappeared when they met face-to-face. According to our information, this worker was having some problems with her boyfriend partly because he was not that romantic when they were together. It seems that she was looking for the romance on the network that she lacked in real life. Perhaps their problems will be intensified if she continues meeting people on-line.

There were also straightforward cases of finding partners for sex on Valentine's Day. By logging on to the network some workers would try to meet other netizens who were interested in romance. They would first exchange personal information and then photos and if both parties were interested in meeting, then they would have sex on Valentine's Day in a hotel. It is interesting that all of these stories were hearsay and we did not interview anyone who had had this kind of experience. Perhaps the subject of sex is taboo in Chinese culture and even though the informants may have had such an experience, they would say that they had just heard about it from other people.

This kind of network romance never happened in the past. It was not merely because of the fact that this kind of new communication technology was not available in the past. It is that the traditional norms were still effective and this kind of action would not be sanctioned. For instance, those migrant workers who still adhere to traditional familial norms and values might not be disposed to join this kind of network romance. The issue is complicated and requires further elaboration.

In the past, particularly in the early 1990s, a married couple could scarcely stay together even they worked in the same factory as the factory did not provide any room for them. It was worse when they worked in different factories, or even worse if the factories were not in the same village or town. Despite the fact that they had comparatively freer latitude than they did in their home villages, most of them were leading a wandering life in Guangdong if they were moving from factory to factory, which led to a very unstable family or emotional life. But it is interesting that this group of migrant workers, which we shall call them the first generation of migrant workers, might not have had any emotional problems with their spouse even though they lived separately for a long period of time. Perhaps, in maintaining the marital relationship, they still adhered to the traditional ideal that marriage was not primarily based on emotional attachment (Wolf 1972). What they considered the most important was to achieve together the functional needs of the family (Yuen *et al.* 2004). This first generation of migrant workers usually works very hard for a few years, and returns to their home villages to start a business or to build a new house when they had saved a certain amount of money (Murphy 2002). In recent years, the migrant workers have predominantly been made up of those born after 1980, that is, after the implementation of the policy of reform and opening up; we shall call these the second generation of migrant workers. Traditional family norms have less influence on this generation of workers and it seems that their personal "likes" and "dislikes" have become increasingly more important factors in determining their personal choices. Thus, they are placing less emphasis on traditional moral responsibility than on the pursuit of romance between husband and wife (Yuen *et al.* 2004). Apparently, nowadays it is hard for young married couples or lovers to stay apart. We have heard a lot of cases in which workers who are lovers or who are married come out for a weekend and stay together in a hotel for one or two nights. There are more cases of divorce than ever before among workers who are separated when working in Guangdong. As being married but staying separately seems to have become a problem that will in turn affect the mobility of the workers, it is now common for factories to provide rooms for married couples; in this case, some factories require the couple to produce a marriage certificate while some do not. The intention behind this kind of policy, of course, is to maintain the stability of the workforce in the factories.

In view of this change and the mobile nature of the migrant workers, their emotional problems concerning marriage, love affairs, and sex have come to the surface. These workers have developed their expanded floating networks in Guangdong, which might be able to provide them with emotional support when facing this kind of problem. The network, however, consists also of their kinsmen and if the problem is shared among them, a conflict between the older and younger generations would probably emerge, leading to further frustration and suffering. The on-line service provided by the mobile phone operators for meeting new people in cyberspace might be a good channel for them to release their emotional problems. This perhaps explains why the on-line chat room service has become increasingly popular. Yet this new service, as mentioned above, is helping to further transform social relationships among the second generation of migrant workers in Guangdong.

Conclusion

The rapid and extensive penetration of the mobile phone has significantly changed the social lives of migrant workers. For instance, the mobile phone has changed the consumption patterns of the workers. In the past, workers led thrifty lives in order to contribute to the welfare of their families at home. Nowadays, they spend more money on mobile phones. Although they earn more than before, they send less to their families back home. Their initial reason for buying a mobile phone is to be able to keep in contact their families more conveniently and frequently, but it is ironic that when they are absorbed in perpetual contact, they are less able to materially assist their village-based kin.

Concerning the impact on social relationships, some studies have maintained that mobile phones enable people to link up the absent present (Gergen 2003) and maintain existing social relationships in expanded spatio-temporal contexts (Pertierra *et al.* 2003). Understanding the case of the migrant workers in these terms, the mobile phone has been conducive to complicated phenomena. When young workers decide to seek their fortunes in Guangdong, they are dreaming about having more freedom from the constraints of tradition. Yet the unfavourable social conditions of the host society have re-anchored them to the networks of their places of origin. The mobile phone has made it possible for workers to reconnect either face-to-face or through cyberspace with their kinsmen or fellow villagers who are scattered far and wide in Guangdong, or even in other provinces. Wherever these kinsmen and villagers are, traditional village norms existing in the form of a floating cyber network will exercise constraints on them when they are connected with mobile phones, either by SMS or by voice calls. Interestingly enough, the mobile phone, a new kind of communication technology, has re-embedded the migrant workers in the bosom of tradition.

The mobile phone, however, also provides conditions for the displacement of traditions (Giddens 1990). Interviews have shown that migrant workers also enjoy developing virtual connections with strangers or joining any virtual network through the mobile phone. Migrant workers in Guangdong face many constraints and predicaments. For instance, the transient nature of migrant workers has made the relationship as well as emotional lives of young married or courting couples more unstable. Developing virtual relationships in the cyber world might provide them with a release from the problems of the real world. This is because the anonymous nature of the virtual relationships would provide them with the feeling that they are under fewer normative constraints when expressing their emotional problems with other netizens. Put simply, workers can express inner feelings to strangers at the other end of the network that cannot be revealed in a proximate social context. This would introduce a new kind of social relationship and the traditional values would perhaps be displaced through the adoption of this new communication technology.

References

Chan, A. 2001. *China's Workers Under Assault: Exploitation and Abuse in a Globalizing Economy.* Armonk, New York: M. E. Sharpe.

Gergen, K. 2003. "Self and Community in the New Floating Worlds." In *Mobile Democracy: Essays on Society, Self and Politics*, edited by K. Nyiri. Vienna: Passagen Verlag.

Giddens, A. 1990. *The Consequences of Modernity.* Cambridge: Polity Press.

Green, N. 2003. "Community Redefined: Privacy and Accountability." In *Mobile Communication: Essays on Cognition and Community*, edited by K. Nyiri. Vienna: Passagen Verlag.

Katz, J. and M. Aakhus, eds. 2002. *Perpetual Contact: Mobile Communication, Private Talk, Public Performance.* Cambridge: Cambridge University Press.

Kopomaa, T. 2000. *The City in Your Pocket: Birth of the Mobile Information Society.* Helsinki: Gaudeamus.

Law, P. 2003. "Mobile phones and the Social Lives of Migrant Workers in Southern China." Paper Presented in Asia-Europe Conference on Computer Mediated Interactive Communication Technologies, Manila, The Philippines, October 20th–21st, 2003.

Metzger, T. 1998. "The Western Concept of the Civil Society in the Context of Chinese History." Hoover Essays No. 21, Hoover Institution on War, Revolution and Peace. Stanford: Stanford University.

Ministry of Information Industry. 2004. http://www.mii.gov.cn

Murphy, R. 2002. *How Migrant Labor is Changing Rural China.* Cambridge: Cambridge University Press.

Parish, W. L. and M. K. Whyte. 1978. *Village and Family in Contemporary China.* Chicago: University of Chicago Press.

Pertierra, R., *et al.* 2002. *Txt-ing-Selves: Mobile phones and Philippe Modernity.* Manila: De La Salle University Press.

Potter, S. H. and J. M. Potter. 1990. *China's Peasants: The Anthropology of a Revolution.* New York: Cambridge University Press.

Solinger, D. 1999. *Contesting Citizenship in Urban China: Peasant Migrants, the State, and the Logic of the Market.* Berkeley: University of California Press.

Wellman, B. 2000. "Changing Connectivity: A Future History of Y2.03K." *Sociological Research Online* 4. http://www.socresonline.org.uk/4/4/wellman.html

Wolf, M. 1972. *Women and the Family in Rural Taiwan.* Stanford: Stanford University Press.

Yuen, S. P., P. Law, and Y. Ho. 2004. *Marriage, Gender, and Sex in a Contemporary Chinese Village.* New York: M. E. Sharpe.

12 *Sexuality as Public Spectacle: The Transformation of Sex Information and Service in the Age of the Internet*

Garland Liu and Joel Lau

Introduction: Critical Issues Regarding the Impact of the Internet on Sex Information and Service

The intimate connection between societal transformation and technological change has long been noticed in history and is a fascinating subject in academia. Principally, two contrasting views regarding the transformation of society and human life through new technology can be discerned. The first view can be coined "techno-utopia" or "techno-visionary," which is an optimistic view founded on the faith that using technology will better human life, improve our standard of living, alleviate scarcity, eliminate inequality, open up new possibilities, or even construct a new society. The second view can be identified as "techno-dystopia," which is a pessimistic view founded upon the distrust and worry of using technology to enhance human life. The enslavement and the alienation of humanity under technocratic domination, or the apocalyptic scenario arising out of the concern over mankind playing God's role are recurrent themes in literature, science fiction, and scientific discourses.

These contrasting visions have found their latest expressions in the age of the Internet and the debates regarding bio-technology and genetic engineering. In *War of the Worlds*, a highly polemical projection of the dark side of the Internet, cyberspace, and virtual reality, Mark Slouka (1995) argues that the spirit of cyberspace is that the "reality is death." According to Slouka, cyberspace is a multiple assault on the human condition. Cyberspace is an "assault on identity," generating a "springtime for Schizophrenia;" an "assault on place" in which physical place is superseded by virtual space; an "assault on community," whereby human communities are transformed into insect-like hives of information superhighways; and finally an "assault on reality," culminating in a "republic of illusion." Other pessimists add that "information technology has the obvious capacity to concentrate political power, to create new forms of social obfuscation and domination" (Roszak 1986:xii).

Although there has been no polemical opposition to *War of the Worlds* in recent Internet debates, each of these themes does have its contenders or supporters, scattering in recent literature. For example, instead of lamenting the disappearance of community in cyberspace, Wellman *et al.* (1996), Blythe and McGrath (1997), and Harasim (1993) have variously argued that the information highway is connecting people and fostering a new form of community and communication. Instead of viewing the growth of the Internet as enslaving humanity in virtual hives and unreality, or as developing towards newer forms of surveillance and domination (Roszak 1986; Stone 1991), many notice the liberating potential of the Internet to eradicate hierarchy and inequalities. Mele (1999) clearly demonstrated, in the case of the Jervay Place redevelopment in Wilmington, how the Internet can be used by local residents as a tool for collective action and empowerment through the gathering of expert opinions and support online.

On the other hand, pessimists may argue that the empowering and liberating potential of the Internet should not be exaggerated. Quite often, socially privileged groups can carry their status to the Internet, or even perpetuate their positions of advantage through their access to the Internet. Finally, network administrators and experts are possibly new sources of social control, power-relations, and hierarchy in the age of information.

The issues of identity deception, fragmentation, and Schizophrenia of the self, and stereotypical and prejudicial interactions are ever present in the Internet. Yet optimists may argue that the potentiality for genuine and open interaction in the Internet, which is free of conventional racial and social prejudices, should not be ignored. The constraints of face-to-face interaction, especially the various stigmas regarding racial, gender, social, class, ethnic, age, nationality, and physical attributes that hinder genuine and open communication, can be put aside at least in the initial stage of online interaction. Kollock and Smith (1999:9) noted that the "poverty of signals [our clothes, voices, bodies and gestures signal messages about status, power, and group membership] is both a limitation and a resource ... individuals will be judged by the merit of their ideas, rather than by their gender, race, class or age." In other words, online communities and communication, despite the possibility of deceptions, also allow for the possibility of a freer and more genuine expression of one's identity and opinions.

Table 1. Possible Manifestations and Tendencies of Four Critical Issues

Critical Issues/ Aspects	Manifestations and tendencies	
1. Community issue	1a. Community destruction	1b. Community formation
2. Social hierarchy and inequalities issue	2a. Enslaving condition: with new forms of inequalities and sustaining/renewing older forms of inequalities	2b. Liberating condition: with greater equality, democracy, and empowerment of disadvantageous groups
3. Identity issue	3a. Identity deception/ concealment: uninhibited, irresponsible expression and destructive communication	3b. Identity articulation: freedom of expression of opinions and genuine communication
4. Reality Issue	4a. Unreality: self-perpetuating and self-contained delusional hive undermining capability to live in the "real world"	4b. Reality: informational treasure mine enhancing "real-world" interaction and survival

Finally, while pessimists may regard the cyber community as delusional and as a "republic of illusion," optimists suggest that online communities, in the sense of social networks, meet any reasonable definition and practice of community (Wellman and Gulia 1999). Kollock and Smith have noted that we "also see the interplay between online communities and the 'real' world. The Internet has been used as an extremely effective tool for carrying out social protest .... The Internet is used as a tool for change in a disadvantageous community, demonstrating a way to route around unhelpful government agencies to find people with information and expertise that they are willing to share" (Kollock and Smith 1999:23).

To sum up, four major issues or aspects with opposing manifestations and tendencies[1] can be identified from the major debates on the Internet, as shown in Table 1.

As shown above, the "techno-dystopian" vision foresees the consequences of 1a, 2a, 3a, and 4a, while the "techno-utopian" vision embraces those of 1b, 2b, 3b, and 4b. But we contend that the age of information is full of possibilities, and we have found no compelling reason to take sides with either of these views. For example, the self-perpetuating property (under the "reality issue")

[1]These four aspects are not meant to be an exhaustive categorization and, in principle, each of these four aspects could be further subdivided into even finer analytical categories. For the sake of discussion, in this paper we will treat these various aspects on the use of the Internet (1–4) as analytically distinct dimensions.

might not be delusional and does not need to be viewed negatively. In different online communities, the major manifestations and tendencies of these various aspects may vary considerably. There is no need to completely buy either the "techno-utopian" view or the "techno-dystopian" view. The central concern of this paper is to use one particular service sector of the sex industry, namely "one-woman brothels" as a case study to examine which combination of manifestations and tendencies best describes the situation in Hong Kong. We found that regarding the provision of sex service information on the web, the different parties involved, namely, sex workers, sex customers, and web information providers, are making use of the emergent information technology very differently. The overall impact of the Internet on the sex service world is a more complex matter than has been envisioned in the above contrasting views.

In what follows, we will first provide some general discussion of the provision of sex services in Hong Kong. We will then move on to compare and contrast the presentation and dissemination of Hong Kong sex service information in the conventional printed mode and in the Internet. Through this comparison, the unique and emergent features of the Internet can be highlighted. We shall then carefully examine the concrete and diverse uses of the Internet by sex customers and sex workers, via the mediation of sex information providers, to analyze which of the abovementioned manifestations and tendencies are prevalent with regard to the sex service world of one-woman brothels in Hong Kong.

Background: Sex Service Sector and One-Woman Brothels in Hong Kong

Sex work covers a wide range of practices for the exchange of sex and sexually related goods and services for financial gain. In this article, we focus on one sector of the provision of sexual services with a "human touch,"[2] one-woman-brothels, rather than with the provision of goods such as erotic garments, sex toys, condoms, medicine for the enhancement of sexual performance, and printed or audio-visual pornographic materials. The sex service sector we explored in the local Hong Kong context mainly involved sexual transactions between female sex workers and male sex customers.

[2]The service sector with a "human touch" in fact contains a whole range of activities from phone sex (only the voice of the service worker over the phone), web-jockeying (the image and the voice of the service worker on screen), erotic shows (the body being physically present but with no or limited physical contact with customers, who are spectators), to various forms of sex services with more direct physical and sexual contact between customers and sex service workers. In a nutshell, the sex service sector involves different forms of services with various levels of physical and sexual contact.

The "one-woman brothel" is the only type of legal sex service operation in Hong Kong. Prostitution in Hong Kong is not illegal, but any facilitating activities that make it possible have been criminalized. Until recently, "one-woman brothels" have been operated mainly by women in their 30s or 40s,[3] who are slipping down the "career ladder" of sex work because of their age or outlook (*Express Weekly*, April 6th, 2004). Although these women may have already passed their prime for sex work and thus are not at the upper end of the market like call girls, escort agents, nightclub girls, or TV/movie starlets, as single operators they often have more control over their work in terms of who, what, when, and where to serve (Kong and Ziteng 2003).

Lately, the number of one-woman brothels in Hong Kong has increased, with the entrance of many younger and physically more attractive women. There are two reasons for this rejuvenation: first, a shift in law-enforcement policy and second, an increasing supply of sex workers from mainland China. Let us elaborate on the two reasons.

Mongkok in the Kowloon Peninsula has long been the area where most "brothels" operate. These brothels have been organized under the "masquerade" of hourly rated villas/motels known as "stables," mostly managed by the triads with their supply of girls coming from Hong Kong, and countries such as Malaysia, Thailand, Japan, Korea, the Philippines, Columbia, Russia, and so forth. Since the mid-1990s, mainland Chinese women have come to dominate the scene. Sex workers in "stables" are nicknamed "horses," with "keepers" (pimps) who send them to villas for sexual transactions. Though illegal, these stables had been tolerated until recently. Only occasionally would police raids be conducted to keep them in line, such as to curb the use of underage girls or missing persons as prostitutes, or forced prostitution.

However, since the beginning of 2004, the police have been conducting raids, breaking up the stables frequently, and disrupting businesses in the area. The reason for the change in law enforcement is that the area has been undergoing re-development and the police have to clear the way for the opening of a five-star hotel and some up-market shopping malls. This is making it very

[3]The impression of a middle-aged Hong Kong woman working in a one-woman brothel was gained from films and through conversations with the staff of an NGO that serves sex workers, during the time I was involved with the NGO. An exploratory study by Kong and Ziteng (2003) of workers from one-woman brothels found that 84.6% of those who answered the questionnaire were in their 30s to 40s, although it cannot be confirmed whether the sample of workers was biased for some reason. In any case, their findings did not seem to contradict a recent newspaper report by the *Express Weekly* (April 6th, 2004) that workers working in one-woman brothels were mainly "middle girls" from Hong Kong.

difficult for the "stables" to continue to operate in the same way as before. Instead of having a few "horses," i.e., women "racing" around villas in the area, the pimps now station the women in small housing units within the area "pretending" to be one-woman brothels. Some of them have even shifted their women to Jordan and Tsimshatsui and put them under the category of "hotel appointment," which means that they no longer have "stables" as bases for business, but need to shift around various three or four-star hotels in Jordan and Tsimshatsui for sexual transactions. As a result, there has been a "spill-over" effect, boosting the number of one-woman brothels, whether within fixed venues of small apartments or moving around hotels. Therefore, in this paper, one-woman brothels refer to both sex service operations managed by individuals and those organized by criminal networks. The masquerading of stables as one-woman brothels serves to conceal the organized nature of sex work from the detection of law enforcement officers.

The second reason for the regeneration of one-woman brothels has to do with a constant supply of women from mainland China since 1990 (Whitehead and Vittachi 1997). This influx of mainland Chinese women for sex services to places such as Hong Kong and Macau, is partially a result of the greater degree of population mobility brought about by the economic reforms in China since the late 1970s. The relaxation of the household registration system has released a much-needed labor force for economic development. A large number of young people have left their villages from the poorer rural areas to look for jobs in the cities, especially in the Pearl River Delta region. Many of these young people work in factories under very harsh conditions. As they are not registered as local residents in those cities; they are not entitled to receiving social welfare from their places of work. Suffering from hardship in the cities, and being migrants and thus far removed from the social control of their original social networks in villages, some of the women are attracted to work in the thriving entertainment business. As China became more open economically in the 1980s and 1990s, there has also been an increase in traffic between the mainland and Hong Kong. Consequently, an increasing number of Chinese women, labeled "northern girls," have come to Hong Kong to take part in sex work. Since the mid-1990s, they have begun to dominate the local sex market (Emerton 2001).

Mainland Chinese women on two-way permits are highly mobile, as they have to leave Hong Kong once their permits expire. They either return to China or go to other places for a short period of time before they can apply to come to Hong Kong again. As a result, there has been a high turnover of mainland Chinese women for the local sex service industry. This has whetted the appetite of customers, most of whom have a taste for "newbies" who claim to be young and fresh. In fact, the influx of young Chinese women largely explains the lowering of the ages of workers in one-woman brothels.

We have noticed that sex service information is most plentiful for one-woman brothels in newspapers and on the web. Undoubtedly, sex service information is the lifeline for this group of workers because their businesses depend very much on the dissemination of basic "path-finding" information, such as when, where, and what they serve. This information also serves as a sort of "stimulant" to boost the demand of customers for sex services. We found that web information for sex workers in one-woman brothels tends to be much more individualized than those for massage parlors or nightclubs in the sense that the advertisements emphasize certain physical, psychological, or sexual attributes of the workers in an attempt to make them stand out from the rest. The constant need to promptly promote the mobile "northern girls" is another impetus for the proliferation of online advertisements. On the whole, sex service information regarding one-woman brothels in Hong Kong, among other sex services, is the most individualized, abundant, and open to empirical investigation. It is within this context that we explore how the Internet has revolutionized the dissemination of sex service information in the sex service sector of one-woman brothels in Hong Kong, affecting both sex customers and sex workers.

Research Methods

Our initial plan was to compare sex service information in the conventional mode and the new Internet mode to explore the impact of technological development on the sex industry. A survey of the conventional printed mode in December 2003 revealed that only a few local newspapers published information on sex services in the form of advertisements (*Oriental Daily*, *The Sun*, *Apple Daily*, and *Hong Kong News Daily*). Among the above newspapers, the Oriental Daily and The Sun dominated the information for operations related to private nursing care/medical treatment, which are mostly one-woman brothels "in disguise."[4]

For the Internet mode, we made attempts to identify major websites for sex services in Hong Kong, utilizing the search engines Yahoo and Google.

[4]We gave up examining magazines and other newspapers, namely *Apple Daily*, because they concentrated on information for massage parlors, saunas, and nightclubs; and *Hong Kong News Daily* because it contains only a very small number of advertisements.

We employed keyword searches, such as "sex service in Hong Kong" (in English and Chinese). However, this way of searching for websites did not seem to be very efficient, as it also brought up many websites for sex-related products, and individual webpages about sexual experiences and viewpoints, which were not our concern. A more effective way of identifying relevant websites was to directly explore the web-links organized by the informational webpage of "Yahoo" (Business & Economy: Company: Adult Information: Adult Services [in Chinese]) and "Timway" (Home: Life: Adult: Services). After some initial surfing, we found that the number of advertisements for "one-woman brothels" was largest of the various types of sex services. We identified over 40 websites specializing in one-woman brothels. A number of them had expired, were under construction, or were duplicates of each other. In the end, we narrowed down the number to 21 websites for further intensive study.

Data for sex service information in the conventional mode was collected through an intensive content analysis of advertisements for one-woman brothels on January 5th, 6th, 11th, 2004; June 24th, 26th, 27th, 2004; and July 1st, 2004. Dozens of telephone calls were made to confirm information on the newspaper advertisements and to check whether the establishments concerned were one-woman brothels. This enabled us to estimate the number of sex service workers who relied on the conventional mode to disseminate information, as well as to compare and highlight the increasing significance of the Internet mode in sex service information. Also by closely studying all of the advertisements for one-woman brothels, we identified three different categories of sex service information and the sorts of limitations to this type of information. All of these data helped us to appreciate the differences between the conventional mode and the Internet mode, and to examine the critical issues regarding the technological transformation of social life.

For the Internet mode, we started off with the tedious task of extensively and intensively surfing the web from January to June 2004. At the most basic level, we gathered data on the number of workers at one-woman brothels who made use of the Internet mode. Through a further in-depth content analysis, we identified certain interesting and emerging features that signified some novel changes in the way sex service information is being disseminated. These features and changes, as our following analyses will make clear, are greatly affecting relationships and interactions in the sex service sector.

Efforts were also made to establish contacts with website administrators and sex service workers of one-woman brothels as informants to countercheck the data collected through the Internet and newspapers. Due to the legally dubious and morally ambiguous nature of sex work in our society, establishing such links was not easy. Through the assistance of workers from two NGOs, some initial contact with the above two categories of informants was made. So far, an initial meeting for about two hours with one sex service worker on September 15th, 2004 was achieved. In this meeting, further information was collected regarding the general situation of one-woman brothels in Hong Kong, the use of the Internet by workers, and effectiveness of the Internet mode in disseminating information about sex services.

Discussion and Findings

Conventional Sex Service Information

Legal and Moral Ambiguities and the Restrictiveness in the Conventional Printed Mode

The function of sex service information is not just to provide details of services available within the sex industry, but also to attempt to stimulate sexual desire, generate demand, and thereby lubricate trade within the sex industry. Traditionally the dissemination of sex service information was mostly initiated by sex workers through the printed media in the form of advertisements, and sometimes by promotional "visit reports" in newspapers and pornographic magazines.

The presentation of information through the conventional means of the printed media is therefore a one-way communication of rather standardized information initiated mainly by the sex service sector through commercial newspaper media to potential customers. Feedback from customers regarding the quality of the services received is rare, if available at all. Even with the popularity of digitalized information technology, e.g., VCDs containing multi-media content distributed alongside printed materials, the overall situation has not changed much. This expansion in the form of presentation alongside printed materials, while making the presentation of information more vivid, has not helped to overcome the "restrictiveness" in the process of generating, distributing, and consuming information on sex services.

This restrictiveness is a combined consequence of the legal dubiety and moral ambiguity inherent in the way the sex industry is perceived by society. The commercialization of sex services is legally dubious and morally frowned upon. In Hong Kong, the act of prostitution is legal, but all activities facilitating prostitution are made criminal. Those involved can be charged with committing a criminal offence. Publishing sex service information to openly promote "prostitution" would have grave consequences for the owners of the newspapers and magazines. Therefore, information in the newspapers on sex services has to be disguised under the categories of "nursing care" or "medical treatment." Even without legal sanctions, the moral sanctions are strong enough to affect retailing practices. For example, even though most newspaper stands and retailers in Hong Kong sell pornographic materials, such materials tend to be placed in an inconspicuous corner of the shop and/or only displayed at certain hours of the day. Sex service workers also face restrictions when placing their advertisements. They cannot turn to any newspapers to promote their services, as property developers do. In Hong Kong, information on one-woman brothels is mainly published by two popular newspapers belonging to the same news corporation, *The Oriental Daily* and *The Sun*.

This restrictiveness also extends to the realm of everyday life. Despite the legality of one-woman brothels, both customers and workers would not like to be identified and made known to the public. Such an identification would probably attract the attention of law enforcers, who would check up on their activities. This would bring upon themselves the moral judgment or social sanctions from their friends, family, and the wider society. More often than not, these moral judgments and social sanctions would cause those involved in the sex industry extreme embarrassment and stigmatization. As such, interactions among various parties in the industry, namely service workers, information providers, and customers, are not and cannot be conducted in an open manner, compared with other more "respectable" trades.

The phenomenal development of information technology in the past decade, however, has been enriching not only the quality of the sex service information that is being presented but also fundamentally altering the relationship of the parties involved in the sex service sector. To lay a foundation for comparison with the new mode of the Internet and examine the impact on the sex service sector, let us first turn to a more in-depth discussion of the features of sex service information in the conventional mode.

Three Categories of Sex Service Information and Their Features in the Conventional Mode

It costs a sex service worker HKD 80 per day to place a 1 cm × 2.5 cm advertisement in the *Oriental Daily* and HKD 25 in *The Sun*. Within the limit of the above advertisement space, 4 large Chinese words and 31 small words are allowed. Thus, the information included has to be compact and economical. For more flexibility, some sex workers prefer to pay twice as much to buy an advertisement space of 2 cm × 2.5 cm. Within the very limited space, each service worker has to decide on what information to provide to her advantage. There does not seem to be any particular formula as to what should be included. Our analysis of the newspaper advertisements identified three categories of information.

• Information on Accessibility

As a rule, the sex worker has to provide at least some information to make them accessible to potential customers. The first category of information is related to details of how to contact the sex service worker. Such information usually includes a telephone number and/or an address, as well as bigger words indicating the regions in which they work. Sometimes, instructions are provided, such as how to get to the address by highlighting nearby geographical features. In our study, we found that the contact telephone numbers that were provided were mostly fixed residential lines rather than cellular phone numbers. This information on access is likely to be more eye-catching and to occupy a substantial proportion of the advertisement space. One last type of such information is about "business hours." Most one-woman brothels open twelve hours a day, with a few operating around the clock.

• Information on Services

The second type of information is related to the kinds of services provided and the fees charged for such services. Due to the issue of legality discussed earlier, it is a peculiar feature of almost all newspaper advertisements for one-woman brothels that the provision of massages is mentioned as part of their services. As a minimum, at least two Chinese characters are used to indicate that the advertisement can be categorized under "nursing care" (in *The Sun*) or "medical treatment" (in the *Oriental Daily*). This arrangement to information on "disguise" sex services under the umbrella of "nursing care" or "medical treatment" is specific for newspapers that are intent on avoiding legal sanctions on activities surrounding "prostitution." The newspapers cannot openly publish information for sex services but are able to "disguise" such information under other advertising categories.[5] Another consequence of legal sanctions is the

[5]It is interesting to note that the news media can avoid getting into trouble with the law by putting the advertisements under the categories of "nursing care" or "medical treatment" while emphasizing the provision of massage services can get the women into trouble. This is because it is illegal to perform massage services in Hong Kong without a license.

codification of information regarding the types of services available. Information on sex services is made elusive and vague for "outsiders," thus making it difficult for the authorities to impose legal sanctions. For example, "oil" stands for the application of massage oil relating to the caressing of the body, "press" stands for massages relating likely to foreplay, and "burst" stands for ejaculation referring to the provision of sexual services. There are also specific kinds of jargon for different types of sexual activities; for example, "mouth burst," "throat deep," and "ice-fire" stand for oral sex; "back garden" and "poisonous dragon screw" stand for anal sex, and so on.

For the same reason, fees for sexual services are made vague. If information on fees is presented at all, it is usually be in the form of a numerical figure such as "150," which means HKD 150 per session. Our survey of the newspaper advertisements revealed that the listed fees for one-woman brothels range from about HKD 100 to HKD 320 per session, depending on the business areas, types of sex services, as well as certain personal attributes of the workers. It should be noted that the pricing information is sometimes elusive and vague. For instance, it is not particularly clear how long one session lasts (one hour or 45 minutes), and what services (BM, BJ, HJ, ML, SM, etc.) are included in the session.

- Personal Attributes

The third category of information is related to the personal attributes of the sex service workers. This category of information is meant to arouse the interest of potential customers for sex service workers. As such, the information provided in this category is highly selective, and only those features thought to be sexually appealing are highlighted. In other words, only those personal attributes that can improve the marketability of sex service workers are included. This information reflects the sex service workers' perceptions of what makes them appealing to potential customers.

There are six sub-categories of information under personal attributes, namely, physical features, age, places of origin, occupational role, personalities, and name. The personal attributes that are most likely to be brought up are physical features such as bust sizes, color of complexion, or specific parts of the body. Sex service workers tend to describe themselves as pretty, busty, and fair with a curvy figure (sometimes a small waist, or being tall with long legs, etc.). In many cases, specific bust (or cup) sizes (ranging from 36 to 42 inches) are given in the advertisement.

As far as age is concerned, most workers describe themselves as young or youthful and some even specify their age (usually from 18 to 22). Occasionally, there are individual workers who describe themselves as "middle girls." These are likely to be women in their thirties or even older, and such a specification means that they cater to customers who prefer older women.

Place of origin, which serves as some kind of differentiation, can be another selling point. Since the 1990s, apart from local women and a handful of women from Taiwan, the Philippines, Thailand, Japan, and Korea, an increasing number of mainland workers have been coming to Hong Kong. Many sex service workers at one-woman brothels give some indication of whether they are "northern girls" (mainland visitors or new mainland immigrants to Hong Kong), locals, or of other ethnic origins. Among the "northern girls," some would specify the provinces or cities they come from such as Sichuan, Nanjing, Suzhou, Hangzhou, Harbin, Shanghai, Beijing, and so on. These labels can help to differentiate them to attract customers with different tastes: some want locals because they can communicate in Cantonese and some prefer women from outside Hong Kong because they are more exotic. Also, it is generally believed that "northern girls" are cheaper as well as more "naive" and eager to serve, and that some places, such as Suzhou and Hangzhou, produce pretty women.

The next sub-category is related to the "occupational role" of the workers. Sometimes, some sex service workers claim to be housewives, kindergarten teachers, air-hostesses, nurses, secretaries, office ladies (OL), or models, who work part time for some extra cash. This is a strategy to satisfy male sexual fantasies relating to certain occupational roles or the sorts of qualities that are attached to people from these occupations. For example, housewives are expected to provide a domestic feel, kindergarten teachers and nurses are thought to be caring and "pure," models and air-hostesses are thought to be chic and sophisticated, and students are thought to be green and fresh, and so forth.

Personalities are also considered worth mentioning by some sex service workers. They often use such descriptions as hot, sexy, fun-loving, lusty, and naughty to project the stereotypical image of the "whore" or, less often, as feminine, gentle, sweet, soft, and pure to project the stereotypical image of the "Madonna," again for the purpose of attracting customers of different tastes. Finally, names are least often included and thus the least important element in newspaper advertisement for one-woman brothels. Only very occasionally will the name of a sex service worker be given in this type of almost faceless one-way communication.

Bluff and Deception and One-way Communication

It is interesting to note that the information category of personal attributes can open up plenty of scope for bluffing and deception. One important point is that the descriptions that the sex workers provide of themselves on the newspapers do not necessarily have to be realistic. Without a face, a body to show and sometimes even without a name, sex service workers can make use of the limited advertisement space as their canvas to put in whatever descriptions that capture the sexual imagination of potential customers with the intention to attract business. Moreover, the social stigma of being a sex service worker is so great that the worker would want to do her utmost to conceal her involvement in the sex service industry from most people, including neighbors, friends, and family members. Therefore, at least from the standpoint of the sex workers, there is no reason to provide genuine descriptions of themselves that may reveal their true identity.

For the above reasons, the depiction of sex workers on newspaper advertisements can be very different from the reality. However, it is very difficult for customers who rely on newspaper advertisements to verify the information that is provided before they arrive at the doorstep of the sex service workers. In principle, newspaper organizations that receive and publish such advertisements would be in the best position to verify the information. However, it would not be in their commercial interest to reject advertisements simply because the descriptions enclosed are not accurate. With mainly two newspapers of the same news group specializing in information on one-woman brothels, customers can hardly turn to other newspapers or printed media for information. Also, given the moral and legal ambiguity surrounding sex services, it is unlikely that customers/consumers will make any formal complaints against the newspapers to appropriate authorities such as the Consumer Council or the sex service worker for disseminating inaccurate or even bogus information. Apart from taking the sex service information in newspapers with a pinch of salt, there is nothing much a customer can do. As discussed earlier, the dissemination of sex service information in the traditional mode is restricted by moral and legal sanctions that keep all of the parties involved "in the closet" (although to a different extent for each party). The consequence is to make the communication flow static and unilateral, generating sex service information that is hard to verify, and creating plenty of scope for bluff and deception.

Emergent Features in the Internet

From what we have explored and discovered under extensive and intensive web surfing, it is no exaggeration to regard the Internet mode as revolutionizing the presentation of sex service information. This mode of dissemination is unprecedented in terms of the currency of the information, the details and vividness of the content, the involvement of sex service information providers, the proliferation of "peer reviews" (i.e., reports by customers), the availability of guides to novices and to proper "conduct," and the use of other advanced informational technology. All of these emergent and fascinating features are inconceivable under the conventional mode. Before we examine these emergent features, let us first outline some general features of sex information on the web.

Overview of the Internet Mode

Like other adult websites, most of the welcome screens at sex service information websites contain sexually explicit or even seductive female pictures, which are used to attract spectators to go on exploring the sites. All of these sites claim that web visitors need to be over 18 years of age (the age of adulthood in Hong Kong) to enter. However, none of these sites involve any age confirmation procedures or practices. They completely "trust" their visitors, or more probably, they couldn't care less about the age of their visitors.

In addition, all of these websites provide free-of-charge surfing, with a few offering special services such as mobile phone SMS and PDA support at a low price. Their incomes, apart from a few that are related to organized triad practices, mainly come from charges for placing individual personal advertisements for sex service workers. Therefore, the number of visitors and the ability to promote sex service exchanges are the things that really matter.

- Names of Websites

One interesting feature is that the majority of these websites include "numerals" in their names, most of which were at first quite puzzling to us. Upon further conjecture and reflection, we found that these numerals are all various abbreviations of "one-woman brothels" in different languages and Chinese dialects or sex-related jargon. For example, "141" is the abbreviation of "one floor one [phoenix]" — one-woman brothels in English and Putonghua, "161" is the equivalent of one-woman brothels in Cantonese and Putonghua, "171" is the abbreviated equivalent of "one room one" in Putonghua, while "67," "167," "69," "160," "166," "168," "668," "6699," and the like are various equivalents of "male genitals" and "sexual intercourse" in Cantonese and "Chinglish" (Chinese/Cantonese English) with a lucky feel (6, 8, 9 are lucky numbers in Cantonese as their pronunciations resemble some Chinese characters standing for fortune or luck). All of these suggest that the websites are supposed to cater mainly to local Chinese and occasionally mainland or English-speaking visitors. In fact, among the sex information websites for "one-woman brothels" that were explored, the majority employed these numerals alone or added numerals with other adjectives or nouns, such as sex, happy, info, girl, easy, finder, and so on. In stark contrast to the "masquerade" orientation of the conventional newspaper media, the explicit nature of these website names demonstrates that they have no pretension to be anything other than one-woman brothels and sex service information providers. Legal sanctions thus do not seem to be an important issue for these information providers, where offshore web domains tend to be beyond the reach of local law enforcement practices.

- Business Booming

As noted earlier, the one-woman brothel sector has been flourishing in Hong Kong in recent years, principally owing to the large supply of younger and "fresher" women from mainland China, who are either new migrants or are visitors on two-way permits. Some of the one-woman brothels, especially those operated by two-way permit visitors, are likely to be organized by triads and some may also well be "stables" in disguise (*Express Weekly*, April 6th, 2004). This expanding sector of "one-woman brothels" is operating today in an extremely competitive market. Within this context, there is an enormous need to provide information to promote the sales of sex services. Given the inexpensive and possibly more effective means of disseminating information, the Internet mode has been thriving lately. It should be noted that about one-third of the websites providing information on one-woman brothels also include information on saunas, massage parlors, "stables," karaoke nightclubs, Internet cafés/bars, escorts, etc., although these are not the subject of our present research.

Except for a few pioneers, the majority of the websites providing sex service information relating to one-woman brothels in Hong Kong were established after 2002. For example, the biggest and the most popular website, in terms of sex worker advertisements and daily online visitors, was established in September 2003, but by late June 2004, it had about 480 advertisements of individual women currently in service, with a total number of accumulated advertisements just shy of four thousand. Such a high accumulated number in just nine months can at least be partially explained by the high turnover of sex workers who hold two-way visitors' permits. The following table categorizes active websites on local one-woman brothels on June 2004 according to the number of personal advertisements of sex workers.

Table 2. The Number of Advertisements for One-woman Brothels by Late June 2004

Number of personal advertisements (June 2004)	Number of websites
1–40	6
41–80	3
81–120	5
121–160	4
161–200	1
201–300	0
301–400	1
401–500	1

In the subsequent discussion, we regard those websites promoting 1–80 active personal advertisements as "small websites," those promoting 81–160 advertisements as "medium websites," and those promoting over 161 advertisements as "large websites."

Through a tedious process of counting all individual personal advertisements put up by currently active sex workers in sex service information websites that were functioning as of June 28th, 2004, in the one-woman brothel sections on "Yahoo" and "Timway," we found the total number of individual personal advertisements to be 2,247, a figure much higher than the range of 450–550 in local newspapers in 2004. We found that only a few sex workers were paying for two or more advertisements on the web. The overall ratio of online advertisements to conventional printed advertisements was about four to one. To sum up, the Internet mode of disseminating sex information is superceding the printed media as the principal source of sex information on one-woman brothels.

- Charges for Advertisements

The charge for placing individual personal advertisements is cheaper in the Internet mode than in the conventional printed mode. From what we have learned in various websites, from journalistic reports in local magazines (for example, *NEXT* magazine), and from conversations with Action for REACH OUT, a local NGO serving sex service workers, the average advertisement fee is around HKD 500 on a monthly basis. Only a few popular websites charge more than HKD 1,000–1,200 per month. Comparing this with newspaper advertisements that charge HKD 80 × 30 (2,400) or HKD 25 × 30 (750) for the smallest slot as mentioned above, advertising on the web is an attractive option.

From the above analysis, we estimate that these websites on one-woman brothels, while charging much less to place advertisements than their newspaper counterparts, can nevertheless obtain advertising income alone of more than a million dollars every month. Given their relatively small operating overheads and lucrative income, it is no wonder that business is booming there. Yet the increasing competitiveness of providing information on the Internet due to the relative ease and low cost of setting up a server for this business also mean that charging an even higher advertisement fee is rather unlikely.

Specific Web Features in Comparison with the Conventional Mode

There are numerous noticeable differences between the conventional printed mode and the Internet mode of sex information. The following discussion will focus on these specific features by comparing them with the conventional mode. These features, as we shall demonstrate, are drastically revolutionizing the one-woman brothel sex service industry. Let us explore these features one by one and then explore their implications on the critical issues of community, social hierarchy and inequalities, identity, and reality.

- Currency of Information

Both the Internet mode and the conventional printed mode are supposed to provide updated information on the one-woman brothel sex services available in Hong Kong. The Internet, however, can provide additional information that the conventional mode cannot. First, in all one-woman brothel websites, new workers will be highlighted on the main page, with details such as their places of origin, locations, names, fees, and dates of arrival clearly presented. Second, those workers who "update" their information, usually after changing their addresses, renovating their venues, or re-entering the market after a break, will also be announced. A typical large website will have around 10–20 new arrivals and updates every day, while a medium website will have a few new entries every day, and a small website will only update its information weekly or monthly. In contrast, under the conventional printed mode, owing to the limitation of space (4 big and 31 small characters) and the poverty of information (a written text alone), there is generally no indication whatsoever about the currency of the information, and of course nothing regarding when the advertisement was first placed. Even though we may see newspaper advertisements highlighting "new entries," this information can hardly be regarded as meaningful. In the newspapers, what is meant by new can range from a few days to weeks or to over months; but on the web, information older than three or four days would not be considered new. Of course, advertisements on the Internet can also be misleading, but as we shall see in a subsequent discussion, there are several mechanisms to validate the truthfulness of the information. Such mechanisms are unavailable in the conventional mode.

The significance of the currency of such information is easy to discern. Many sex service customers are looking for "newbies," i.e., workers who are new to the business. Being new and fresh are highly marketable features. There are numerous reports and discussions in these sites revealing how excited, and sometimes how disappointed, those customers are after have just visited new and fresh workers. So it is a customary practice for all of these websites to place this current information on their main page. The majority of these websites present newbies with miniature erotic photos on their main pages, which are hyper-linked to the pages of individual personal advertisements. Some of these websites that support PDA/PPC version and provide SMS and GPRS functions, also highlight the currency of the information as their major selling point.

- More Detailed and Vivid Information

A typical webpage featuring advertisements for one-woman brothels will contain such detailed information as personal attributes, photographs of women in seductive poses, information on location, and clear-cut descriptions of services and prices. Compared with the conventional printed mode, the Internet mode includes all of the contents that can be found in the printed mode, but in greater detail and vividness. It also includes other information unavailable in the printed mode.

The first category of information in the conventional printed mode regarding accessibility, such as region/address, business hours, and contact phone numbers are all present on the web. In the conventional printed mode this information is provided in the form of written text with an address and/or instructions highlighting nearby geographical features, and is supplemented through contact over the phone. In the Internet mode, the phone, if used at all, is for the confirmation of appointments but not for "path-finding." In any website, there are photographs (or even maps) showing the worker's location, street corner scene, the entrance of a building, the front door of the apartment, the interior, and even the bed where sex services are performed. These apartments or sub-units are air-conditioned, and during winter time, the presence of heaters is also highlighted as a selling point.

The second category of information noted in the printed mode, on the provision of services and fees, is also present on the web, but in greater detail and with novel properties. We will come to them later in the section on "search engines and categorization." It is noted that a very detailed charging scheme for various services is listed on most sex workers' personal information webpages, leaving little room for negotiating fees for services. In one FAQ section of a website, customers are repeatedly advised not to negotiate and bargain with the workers, assuming that they arc fully informed of the fees and thus should be able to decide whether they can afford it beforehand. This contrasts drastically with the elusiveness and vagueness of service information found in newspapers. Service information on the web is more open and transparent in that it is very detailed, far more vividly presented and juicier in content than its printed counterpart.

The third category of the personal attributes of sex workers is a subsidiary if not a "concealed" and "imagined" aspect in the printed mode, but we found it highlighted in the workers' personal advertisements on the web. All of the workers appearing on the web gave their names or nicknames (in Chinese and in English), whereas the majority of workers in the printed mode are nameless. These names can of course be false. It is open knowledge and common practice that if a worker has received very bad comments or visit reports (or after-reports) from web-surfers who are supposedly her customers, she may change her name, move to a new location, place her advertisement in other websites, or any combination thereof, to promote herself once again. The implications of this aspect on the "surveillance" of the performance of services will be discussed later.

Information with regard to other physical attributes, such as height, body figure, "cup-size," age, place of origin/nationalities/languages spoken, and skin complexion are all presented on the web. Coupled with such information are seductive pictures showing the bodies of the workers, ranging from a few erotic mini-pictures to a dozen more, to even short videos in certain websites. Usually this visual information is produced by the information providers and is thus comparatively more reliable than the self-portrayal of sex workers presented in the printed mode. It is noted that most pictures are "copyrighted," with the name of the websites being presented in the corner or somewhere in the pictures. A web-administrator we interviewed revealed that this is a reaction to the unscrupulous practices of a new sex information website that copied and stole pictures from other established websites. In contrast with the conventional mode, this vivid and more reliable graphic information boosts the confidence of Internet visitors that they are more likely to find the types of girls that they want, at least physically speaking. Communication with a sex worker who advertised both on the web and newspapers confirmed that very few customers who obtained information from the web turned away at her doorstep as compared with those who obtained information from newspapers.

• Search Engines and Categorization

Another striking feature in the Internet mode is the categorization, through the aid of information technology, of the sex services provided. In all large and many medium-sized websites, there are powerful search engines that help web surfers, as potential sex customers, look for what they want. The general categories/attributes of workers available for searching usually include the following:

1. Region/Area/District
2. Fee
3. Body figure (bust-size and cup-size)
4. Age
5. Nationality/Place of Origin/Language Spoken
6. Height.

With this search engine, we found that the overwhelming majority of sex workers speak Putonghua, which signifies that they are "northern girls" and confirms their important role in rejuvenating the one-woman-brothel sex service sector. In a few large and medium-sized websites, their search engine includes details of sex services. Web surfers can then select their preferred services and look for workers that perform them. The following items are found in one of the websites with very explicit and elaborate categories of sex and subsidiary services for searching:

1. Make Love
2. Blow Job
3. Hand Job
4. Second Shot
5. Comfort Muscle
6. Bust Shoot
7. Ass Licking
8. Mouth Cum
9. Face Cum
10. Uniform
11. Anal Sex
12. SM

13. Wet BJ
14. Stay Overnight
15. 24 Hours Service
16. Hotel Appointment
17. English Speaking.

In some websites these services are presented graphically with icons, while some others use popular abbreviations such as ML, BJ, HJ, SS, SM, etc. Again, compared with the ambiguity and negotiability of services and fees in the conventional printed mode, sex service information on the web is clear-cut, systematized, searchable, and very explicit, and thus a lot more approachable for users.

Thus far, the provision of sex service information in the Internet may seem to be far less restrained than in the conventional printed mode. The presentation of information does not seem to share the same kinds of restrictions as discussed previously. However, it should be noted that sex service information on the web is subject to the same legal sanctions as in the conventional mode. One possible reason for the explicitness of the latter is the low setup costs and overheads of commercial websites. In the conventional mode, the owners of newspapers have much more to lose by violating the law. Another reason may well be the difficulties of enforcing the law when monitoring the Internet mode of information, and thus of imposing legal sanctions on "offenders."

At this stage, we can see that the Internet mode seems to have many advantages over the printed mode for the parties involved in the sex service industry, with regard to its costs and the quality of the information provided. However, the conventional printed mode has not been totally replaced. There are a few probable answers to the continued existence of the conventional mode. First, the details and vividness of information are a "blessing" for younger and physically more attractive workers, but not as welcoming for older women and those who are physically less attractive. Second, Internet users are mostly younger adults capable of utilizing the latest technology. Yet there are still potential customers who are less equipped. They have difficulties obtaining information on the web and therefore have to turn to the conventional printed media. Third, one-woman brothel advertisements on the web and in the print mode probably cater to two different consumer markets, with the Internet aiming at people who can afford higher prices than those who turn to newspapers for information. The range of sex service fees per session advertised in the newspapers (HKD 100–350) is considerably cheaper than that on the web

(HKD 250–800+). There are probably also other factors, but further research would be needed to uncover them.

- Involvement of Information Provider: Activity Guides, Comments, and Website Maintenance

Information in the newspapers on one-woman brothels are just advertisements. The worker contacts the press, provides information, pays the fee, and the advertisement is placed. There is no further interaction between the information provider (e.g., editors and marketing teams) and their clients (sex workers). The interaction between the information provider and the reader is non-existent (we have never seen any complaints or letters to the editor regarding the provision of information on sex services).

In the Internet mode, the information provider who is also the web administrator, plays a far more active role. In most large and medium-sized websites, there are forum and report sections. In the forum section, information providers set up elements for proper conduct and "rules of the game." As we noted earlier, these websites generate their income through workers placing advertisements in their websites with the expectation of promoting their businesses. It is in the interest of website information providers to attract web surfers/visitors and turn these surfers into the customers of their clients. As a result, these websites are not just trying to present information, but also actively promoting sex services through various activities.

One kind of popular information is a detailed guide for visiting sex workers, clarifying the complicated jargon in the trade, suggesting what is to be expected, what can be done, and how to avoid being regarded as a fool. Many such guides include information about AIDS and other sexually transmitted diseases, and usually urge visitors to be very cautious in their proceedings.

Second, information providers usually make some short written comments on the "personality" and "selling points" of their clients in the latter's advertising webpages. This contrasts with the convention mode, where the workers themselves highlight their attributes and selling points. Such outsider comments, accompanied by photographs, appear to be more "objective" and "reliable." In most advertising webpages, information providers grade the various attributes of the workers. The look, the body, the skill/technique, and the service/attitude are the most common criteria for evaluation.

The above efforts on the web by information providers are already spectacular compared with those seen with the conventional mode, but these pale beside their role in maintaining the enormous flow of information in the websites. In all large and some medium-sized websites, there is an "after-report" section. This "title" is chosen because it is reminiscent of the "after-reports" of local horse racing sessions and the fact that consuming sex services is often compared to horse riding in the local context. In this section, web surfers describe their encounter with sex-workers, very often in vivid and provocative details with some subjective evaluations of the services received. Clearly the privacy and identity concealment grant to sex customers as web surfers has entirely removed the moral sanctions and restrictiveness imposed on the conventional mode of sex service information.

Web information providers, as network administrators of their sites, "censor," edit, and comment upon forum and after-report discussions from time to time. Like other discussion groups and websites, incoming messages are posted before they are read but could be modified if necessary. To handle such an enormous flow of information, especially in a large website, several administrators are needed. These administrators will judge whether or not the after-report is truthful, how far it might hurt the businesses of the sex workers (their clients), whether it is written with malicious intent, and so forth. We found that if a piece of information is misleading, the administrators might still post and maintain it, but will add harsh comments and critiques. In some cases, the administrators might even follow up on the complaints of customers. Upon reading thousands of such after-reports, we estimated that about half of the after-reports are mainly descriptive, around one-third of them express a great deal of appreciation for the services performed by the workers, and less than one-sixth contain harsh remarks from dissatisfied customers.

At first we were surprised that after-reports with harsh remarks about clients (sex workers) are listed at all. Would it not be in the interest of the information provider to attract more advertisements, thereby making them more likely to "appease" the clients by presenting them in as favorable a light as possible? But like other mass media, the web administrators cannot simply ignore the complaints and views of web surfers. If web surfers are not allowed to express their opinions or grievances, they might become alienated and visit other websites for information. Failing to have the support of visitors could have disastrous consequence for a website and lead to a vicious circle. Losing credibility means losing visitors, and in the long run, means that the sex service workers featured in the website cannot take advantage of their advertisements. They might quit advertising at a site and eventually turn to other sites. This explanation is somewhat supported in our subsequent interviews with the web administrators.

For web administrators, a delicate balance, therefore, has to be reached between presenting information that is as favorable to the sex workers as possible and presenting information that is as truthful and uncensored as possible for web surfers. Given the enormous flow of "traffic" in large websites, this is no easy task. Let us illustrate the work involved by examining the situation of a large website. The administrator's guide to the after-report format and after-report writing had been read more than 30,000 times (by the end of June 2004). In a three-month time period (April–June 2004), 1,943 after-reports had been posted with 85,433 "responses." In addition, 2,397 general questions had been asked in the "girl-finding manual," with 50,142 "responses." This brought a total of about 135,000 response entries for administrator(s) in that website to handle, more than 1,500 every day. All of this clearly demonstrates that one-woman brothel web administrators are forming an emergent "stratum" in the provision of sex services. The implications of this striking phenomenon will be further explored later.

- The Emergent Community of Sex Customers as Web Surfers

In the conventional mode of providing information on sex services, there is little room for community formation and discussion. Newspaper advertisements are only one-way, and while the local pornographic magazines we surveyed did have some room for readers to express their opinions, the space to do so was very limited and there were time lags. Adding to the technical problems to free and convenient communication is the moral stigma attached to sex service in general. It is not just sex workers who are frowned upon, but also sex service customers.

With the advent of the sex service information on the web, everything is changing. Sex customers, who used to be secretive consumers, have been "liberated" in their transformation from sex customers into "customers-cum-surfers." The privacy and concealed identity of web-surfers allows them to evade the legal and moral ambiguities inherent in sex service that restrict the expression of opinions. The open platform of the Internet and the privacy has given sex customers a revolutionary channel to express their views, seek out relevant information and activity guides, and eventually to develop a "community" life.

The privacy associated with net surfing at one's own home removes the "restrictiveness" found in the conventional mode. On the web, surfers are free to explore sex information and express their opinions, describe their experiences, and ask for help. We found that in all large and many medium-sized sites, website surfers can join as members, a status which grants them the right to get more (member only) information, to respond to a topic or an after-report and, more importantly, to initiate a topic and an after-report for discussions and responses. Registration for membership is easy and straightforward. A screen name, a private password (for future log-ins), and a functioning email address are all that are required. The site keeps track of the number of postings that members have made and accords them a corresponding prestige, usually from one to several stars. We saw quite a few members with several thousand postings. At some sites, they even restrict certain areas and discussions to "prestige members" with multiple stars.

Fun seeking and extreme vulgarity characterize the general tone and the language of the visitors' after-reports and discussions. As web-surfers, these sex customers are often "uninhibited," doing things that they normally would not do. We saw quite a number of discussions seeking underage (<16) girls for sex, with others who claimed to be girls under 16 openly soliciting sex transactions. In some cases people reported that they caught malicious computer virus upon receiving further information from these supposedly "underaged girls." Quite a number of surfers condemned the act of seeking underaged youngsters as criminal and immoral. In the responses of numerous after-reports, we also saw many cases of surfers with different opinions and value orientations. In one case, a customer complained that he had visited a worker during her menstrual cycle and therefore did not perform sexual intercourse. He felt cheated. Some surfers agreed with him, but others argued that as men they needed to appreciate the hardship of these women who had to work even during their menstrual cycle.

To conclude, the Internet mode of providing sex information creates unprecedented room and "privacy" for sex customers to express their identity in "public" without the fear of being identified and stigmatized. The various discussions and communication available on the web also create enormous social space for the customers themselves, the customers and web administrators, or to a much lesser extent, the customers and sex workers, to interact.

Now let us turn to our final section to consolidate our discussion by re-examining the manifestations and tendencies on one-woman brothel websites with regard to the four critical issues of community, social hierarchy, identity, and reality.

Critical Issues and Implications Regarding the Impact of the Internet on Various Agents in Sex Service Information

From our preceding findings and discussions, we come to the following conclusions. First, the Internet is not only bringing more information at faster speeds and with greater convenience, but also radically transforming the connections and interactions between various agents involved in sex information. The sex service customers of one-woman brothels used to be a scattered group of individuals who usually sought sex services on their own. Apart from close friends with whom they may have shared these practices and chatted privately from time to time, there was no social space or network for them to express their views, seek opinions, or discuss their experiences and/or fantasies. The rise of the Internet has undoubtedly opened up the room and opportunity for them to exchange information and views as information recipients and fellow sex service customers (they address each other as "brothers"). They share with their brothers on the web information about certain sex workers, express their views about the consumption of sex and, at times, their fears as well as knowledge about sexually transmitted diseases. Obviously, a community among sex customers has formed with the help of this new mode of providing sex service information. One of our future plans is to explore the values and worldviews of sex customers, and their characteristic modes of interaction and communication.

Upon further reflection, we also see that the picture may not be as simple as it may first appear. While "community formation" is obvious among sex customers, we seldom saw sex workers participating in the online community. However, we saw a greater connection between web administrators and sex customers over a variety of topics such as activity guides, the legality of sex services, medical concerns, ways of handling various situations, and so forth. This brings us to issues regarding the liberating and empowering potential of the Internet. Sex customers can access more information, seek expert opinions, share their experiences, and express their views fairly easily. In almost every respect, these people are "empowered." The group of sex workers who fail, for whatever reasons, to utilize the technology may become relatively impoverished. If they perform badly once, their poor performance may become known to the "public." Their "private" acts with their customers may be openly discussed and made fun of. This is a big intrusion into their privacy, with a profound asymmetry in power relations. The sex workers, even through using false names, have to make public their body and appearance through their advertisements, but the same is not true for the sex customers who only needed to show their screen names. This situation reveals the inequality of anonymous men being dominant over identifiable women with the latter's absence. The Internet mode of sex service information facilitates the constant "surveillance" of the service performance of the sex workers, turning them into commodities. Under this situation, the Internet obviously works to the advantage of the sex service customers. At this stage of its development, the liberating potential of the Internet is rather discriminatory.

However, we need not come to the conclusion that web surfers are all taking advantage of their anonymity to attack and degrade sex workers. We found a substantial number of responses and comments condemning the rudeness and inhumane practices recorded in the after-reports of some customers. The Internet mode of communication is promoting democratic participation among sex customers, and this democratic element may condemn the socially and morally extreme acts of some sex customers, thereby indirectly working to the advantage of the sex service workers. More research on the communications among sex customers and sex workers is needed to shed more light on issues of social hierarchy and inequality arising from the use of the Internet.

Another potential source of hierarchies and inequalities come from the emergent agents of web administrators who have enormous capacity to select and dominate information on the web. We note, however, that they are subject to the competing demands of treating their clients (the workers) in a favorable manner and being truthful and open to the visitors to their sites (sex service customers). At the present stage of our study, we do not have enough evidence to analyze the potency of their power and actual practices. Further research needs to be conducted on the following issues to assess their power and standing in the one-woman brothel sex service sector: (1) how these websites operate on a day to day basis; (2) how they solicit workers for business; and (3) how they deal with (if they are not themselves) "pimps" of organized brothels that masquerade as one-woman private brothels.

Regarding the issues of identity and reality, we have already noted that the anonymity of sex customers encourages them to express their opinions and experiences. To a great extent, this helps them to articulate their identity as sex service customers or persons enjoying paid sex services, a kind of identity that was almost impossible to express in the past. The availability of information and expert opinions on the web is a treasure mine of information that enhances the "real-world" interactions of sex service customers. Yet the well-known ability to hide one's identity on the web also releases "uninhibited" expressions, such as abusive language, anti-social and hate-filled talk, and "untruthful" expressions, such as misleading and false information. Thus, the issues at stake are: How far does identity deception weaken the credibility of the messages and information on the web? To what extent has sex information on the web created a group of delusional surfers who cannot fare better in their real lives?

From what we have studied so far, there are two mechanisms to counter these uninhibited expressions and false information. First, web administrators could censor, alter, or even ban visitors' postings if they find them to be excessively vulgar and "explicit." Censorship was exercised occasionally, especially when foul language was used or unacceptable topics such as bestial sex were discussed. But there was great leniency in the treatment of fun-seeking dirty talk, so the contents of discussion boards and after-reports could be rather obscene and provocative. Second, obviously fabricated after-reports or the unjustified degradation of workers were banned or were still allowed to be posted but was strongly condemned. Dubious postings often generated critical comments from knowledgeable visitors, so there was little room to sow misleading information. But whether the two mechanisms identified here are enough to eliminate these expressions and false information can only be decided by subsequent research.

We conclude that the growth of the Internet mode is a blessing for sex service customers in a number of aspects, but at most a mixed blessing for sex service workers. We can come to this conclusion only because we have made no prior condemnation of the exchange of sex services. But the public, who frown upon these sexual practices, and law enforcement agents, who are concerned about prostitution, may think otherwise. These issues, however, can only be explored and resolved in future studies regarding public opinion, and the law enforcement issues of the operation of these supposedly one-woman brothels. We hope that our findings and discussions can provide an informative overview of the presentation of sex service information pertaining to one-woman brothels in Hong Kong, articulate the emergent features of sex service information on the web, and shed some light on a few critical issues regarding the co-construction of technology and society in the age of the Internet.

References

Action for REACH OUT. 2003. "Sex Industry in Hong Kong in Flux." October, Hong Kong.

Action for REACH OUT. 2004. "Integrated Services for Migrant Sex Workers". Paper Submitted to *International Conference on Sex Work Rights*, February, Hong Kong.

Blythe, J. and C. McGrath. 1997. "Web Linkages as Social Networks." Paper Presented at Sunbelt International Network for Social Network Analysis Annual Meeting.

Davidson, J. O'C. 1998. *Prostitution, Power and Freedom*. Cambridge: Polity Press in Association with Blackwell Publishers.

Emerton, R. 2001. "Trafficking of Women into Hong Kong for the Purpose of Prostitution: Preliminary Research Findings." Occasional Paper No. 3, February, Centre for Comparative and Public Law, Faculty of Law, University of Hong Kong.

Express Weekly. 2004. "Police Operation to Stamp out Night Businesses." June 6th, pp. 26–28, 30 and 32.

Harasim, L. M. 1993. "Networlds: Networks as Social Spaces." In *Global Networks: Computers and International Communication*, edited by L. M. Harasim. Cambridge, MA: MIT Press.

Hochschild, A. R. 1983. *The Managed Heart: Commercialization of Human Feeling*. Berkeley and Los Angeles, California: University of California Press.

Kong, T. S. K. 2001. "What It Feels Like for a Whore: The Body Politics of Sex Workers in Hong Kong." Paper Presented at the International Conference on Sex Workers: Global Experiences and Local Responses, June.

Kong, T. S. K. and Ziteng. 2003. "A Research Report on the Working Experiences of Hong Kong's Female Sex Workers." Centre for Social Policy Studies, Department of Applied Social Sciences, The Hong Kong Polytechnic University.

Mele, C. 1999. "Cyberspace and Disadvantaged Communities: The Internet as a Tool for Collective Action." In *Communities in Cyberspace*, edited by M. A. Smith and P. Kollock. London: Routledge.

Roszak, T. 1986. *The Cult of Information: The Folklore of Computers and the True Art of Thinking*. New York: Pantheon.

Slouka, M. 1995. *War of the Worlds: Cyberspace and the High-Tech Assault on Reality*. New York: Basic Books.

Smith, M. A. and P. Kollock, eds. 1999. *Communities in Cyberspace*. London: Routledge.

Stone, A. R. 1991. "Will the Real Body Please Stand Up? Boundary Stories about Virtual Cultures." In *Cyberspace: First Steps*, edited by M. Benedikt. Cambridge, MA: MIT Press.

Tsui, C. 1994. "Anna's Dream: A Prostitute in Sham Shui Po Tells Her Story". *Varsity*, February, 9–10.

Wellman, B. *et al.* 1996. "Computer Networks as Social Networks." *Annual Review of Sociology* 22:211–238.

Wellman, B. and M. Gulia. 1999. "Virtual Communities as Communities: Net Surfers don't Ride Alone." In *Communities in Cyberspace*, edited by M. A. Smith and P. Kollock. London: Routledge.

Whitehead, K. and N. Vittachi. 1997. *After Suzie: Sex in South China*. Hong Kong: Chameleon in Association with Corporate Communications.

Notes on the Contributors

Wenhong Chen is a Ph.D. candidate in the Department of Sociology and a research associate at NetLab, the Centre for Urban and Community Studies at the University of Toronto. Her research interests include social networks, economic sociology and the interaction between technology and society. Ms. Chen's dissertation examines how transnational entrepreneurs rely on social networks and the Internet to access and mobilize resources.

Wai-chi Chu is a lecturer in the Department of Applied Social Sciences at The Hong Kong Polytechnic University. He is currently a Ph.D. candidate at Queensland University, Australia. His research interest is in indigenization of social science knowledge. In addition to publishing work relating to methodological issues in China and Hong Kong studies, he is the co-editor of the book *Shehui Kexue Bentuhua: Duoyan Shijiao Jiedu* (2001) (Reflections on the Approaches to the Indigenization of Social Science). His current research concerning new migrant workers and the use of mobile telephony in China is a continuation of his academic pursuit linking up substantive research with meta-theoretical consideration.

Alberta Contarello is Professor of Social Psychology at the University of Padova, Italy. Her current interests relate to the social construction of knowledge — with particular regard to new technologies — social psychology and qualitative research, social psychology and literary texts. Among her recent publications on the theme of ICTs are:

- 2003 Contarello, A. "Body to Body: Co-presence in Communication." In *Mediating the Human Body. Technology, Communication and Fashion*, ed. L. Fortunati, J. Katz, and R. Riccini.
- Contarello, A. e Sarrica, M. (in press) (available on-line 21 September 2005) "ICTs, Social Thinking and Subjective Well-being: The Internet and its Representations in Everyday life." *Computers in Human Behaviour.*

Leopoldina Fortunati teaches Sociology of Communication and Sociology of Cultural Processes in the Faculty of Education at the University of Udine. She has conducted several researches in the field of gender studies, cultural processes and communication technologies. She is the author of many books and was the editor of *Mediating the Human Body: Technology, Communication and Fashion* (2003) with J. Katz and R. Riccini. She has published many articles in important journals. She represents Italy on the COST Technical Committee for Social Sciences and Humanities and on the action COST A20 "The Impact of Internet on the Mass Media in Europe". She was involved in the European research project SIGIS "Strategies of Inclusion: Gender and the Information Society" and COST248 "The Future European Telecommunications User" and was vice-chairperson of COST269 "User Aspects of ICTs". She is the co-chair of the International Association "The Society for the Social Study of Mobile Communication" (SSSMC) which aims to facilitate the international advancement of cross-disciplinary mobile communication studies. She has organized several international workshops and conferences. Her works have been published in eight languages.

Wei Hong is a Ph.D. candidate in the Department of Sociology at the University of Illinois, Chicago. She studies innovation and the sociology of science. Her masters thesis examined the sources of authority in a Chinese lab setting. Her dissertation research was concerned with the geographic dimension of knowledge flow from university to industry in China. She is also co-authoring, with Professor John Walsh, a paper addressing the secretive behavior of scientists across disciplines and over time; a brief version of which was published in *Nature*.

Joel Lau obtained his M.Phil. degree from the University of Hong Kong and is pursuing his doctorate at the University of Toronto. He has been lecturing sociology courses in various Hong Kong universities and in adult continuing education since 1989, covering subjects from introductory sociology, sociology of education and environmental sociology to Hong Kong society and Chinese family and culture. He has developed research interest in topics including social and political movement, modern environmentalism and residential preference, and sexuality.

Pui-lam Law received his Ph.D. in sociology from The University of New South Wales, Australia, and is currently Assistant Professor in the Department of Applied Social Sciences at The Hong Kong Polytechnic University. His research interest is modernity and rural China. He co-edited *Dangdai Zhongguo Nongcun Yanjiu (Shang): Lilun Tansuo* (Studies on Contemporary Rural China, vol. 1, Theoretical Exploration) (2000) and *Dangdai Zhongguo*

Nongcun Yanjiu (Xia): Shizheng Diaocha (Studies on Contemporary Rural China, vol. 2, Empirical Investigation (2000). He also co-authored *Marriage, Gender, and Sex in a Contemporary Chinese Village* (2004). Recently, he has been working on mobile telephony and migrant workers in southern China.

Garland Liu obtained her doctorate degree in sociology from The University of Aberdeen, Scotland. Her Ph.D. study mainly looked into how Chinese people managed to make a living through the economic enclave of the catering trade with distinctive ethnic characteristics in Britain. Upon returning to Hong Kong, she worked briefly with The Chinese University of Hong Kong and The University of Hong Kong. In 1995, she joined The Open Learning Institute of Hong Kong which was renamed The Open University of Hong Kong (OUHK) in 1997. She is now Assistant Professor there in the School of Arts and Social Sciences responsible for distance education courses in both the Social Sciences and the Law Enforcement and Security Management Programmes. She also teaches courses in introductory sociology and Hong Kong society in the OUHK's full-time face-to-face programs. During her time at the OUHK, she has developed research interest in topics such as policing and law enforcement as well as gender and sexuality.

Luca Giuliano is Associate Professor of Sociology at the University of Rome, "La Sapienza", Faculty of Statistical Sciences, Department of National Accounts and Social Processes Analysis, where he teaches methodology of social sciences and textual data analysis. He is the author of *La logica della scoperta nelle scienze sociali* (2003) and *Parole in libertà. Analisi statistica e linguistica dei discorsi di Berlusconi* (2006).

Yinni Peng obtained her Bachelors and Masters Degrees in Sociology at Peking University, PRC. She is currently a Ph.D. student in the Department of Sociology at the Chinese University of Hong Kong, studying ICTs and migrant workers in southern China.

Raul Pertierra obtained his Ph.D. in anthropology from Macquarie University, Sydney, Australia. He taught at several universities in Australia and elsewhere, but mostly in the School of Sociology at The University of New South Wales, Sydney, Australia, where he is presently a Visiting Fellow. Dr. Pertierra is a Visiting Professor in the Department of Sociology and Anthropology, Ateneo de Manila University. His most recent publications include: *Txt-ting Selves: Cellphones and Philippine Modernity* (2002), *The Work of Culture* (2002), *Mobile Phones, Identity and Discursive Intimacy in Human Technology* (2005).

Lydia Sciriha obtained her first degree in languages from the University of Malta. As a Commonwealth scholar she followed postgraduate studies at the University of Victoria in British Columbia, Canada. Since 1987 she has been teaching sociolinguistics and related areas at the University of Malta. She has also lectured at the University of Cyprus and at Victoria University in Melbourne, Australia. Besides a number of articles in scientific journals, Professor Sciriha is the author of a number of books. She is also the editor of *Humanitas: Journal of the Faculty of Arts.*

Raimondo Strassoldo has been Professor of Sociology at the universities of Trento, Trieste, Milan and Palermo. He has also been a DAAD scholar at the Technical University, Munich. His research interests have ranged over time from the sociology of space (rural, urban, territory, environment) to the sociology of international and inter-ethnic relations. On these topics he has written extensively, mostly in Italian. Among his publications that we can cite are: "The Social Construction and Social Analysis of Space," in B. Hamm and B. Jalowiechi (eds.) (1990) *The Social Nature of Space.* Warsaw: Panstowe Wydawnictwo Naukowe; "Ethnic Regionalism versus the State: The Case of Italy's Northern Leagues," in L.O'Dowd and T.M.Wilson (eds.) (1996) *Borders, Nations and States.* Aldershot: Averbury. Since 1994 he has taught Sociology of Culture and the Arts at the University of Udine, Italy.

William Wai-lim Wong is Assistant Professor in the Department of Social Work at the Hong Kong Baptist University. He co-edited *Transformation of Policy and Services for Older People* (2005). He has also contributed a book chapter on the diversity of ageing theory and personal economy of ageing to this collection. His research interest is mainly on ageing and ageing policy in Hong Kong. He is currently researching the provision of elderly residential care in Hong Kong.

Kerstin Wüstner is currently Professor of Applied Psychology in the Faculty of Philosophy and Social Sciences at the University of Augsburg, Germany. One of her research fields is the analysis of psychosocial aspects of human genetics. In a recent research project, which was carried out together with colleagues in Australia and Japan, she compared attitudes towards human genetics in different countries. Important publications related to this field are:

- 2003 "Ética profesional. Los genetistas humanos en Alemania." In *Ética profesional e identidad institucional*, ed. Ana Hirsch Adler. Centro Estatal de Investigación y Desarrollo Educativo de Sinaloa (CEIDES), Mexiko.
- 2003 "Ethics and Practice: Two Worlds? The Example of Genetic Counselling." *New Genetics and Society* 22(1): 61–87.

Shanhua Yang is currently Professor in the Department of Sociology at the Peking University. He publishes extensively on rural Chinese society.

Index